Fourth Industrial Revolution and Business Dynamics

Nasser Rashad Al Mawali ·
Anis Moosa Al Lawati ·
Ananda S
Editors

Fourth Industrial Revolution and Business Dynamics

Issues and Implications

Editors
Nasser Rashad Al Mawali
College of Banking
and Financial Studies
Muscat, Oman

Anis Moosa Al Lawati
College of Banking
and Financial Studies
Muscat, Oman

Ananda S
Postgraduate Studies
and Research Department
College of Banking and Financial
Studies
Muscat, Oman

ISBN 978-981-16-3249-5   ISBN 978-981-16-3250-1   (eBook)
https://doi.org/10.1007/978-981-16-3250-1

This Palgrave Macmillan imprint is published by the registered company Springer Nature
Singapore Pte Ltd.
The registered company address is: 152 Beach Road, #21-01/04 Gateway East, Singapore
189721, Singapore

FOREWORD

Fourth Industrial Revolution (4IR) is a fusion of advances in artificial intelligence (AI), robotics, the Internet of Things (IoT), 3D printing, genetic engineering, quantum computing and other technologies. It is the collective force behind many products and services that are becoming indispensable in recent times. Consequently, the Industry 4.0 is paving the way for transformative changes and disrupting almost every business sector in the economy. The 4IR has the potential to raise global income levels and improve the quality of life for people across the globe. On the contrary, economists Erik Brynjolfsson and Andrew McAfee have pointed out that the revolution could yield greater inequality and may disrupt labour markets. The 4IR led automation may displace workers with machines and lead to increase in the gap between returns on capital and labour. On the other hand, there is also a possibility of net increase in safe and rewarding jobs. At this point, it is difficult to foresee and predict the scenario that would likely to emerge and impact on economy and human life. Hence, there is a need for new body of knowledge in this area for the stakeholder to make a meaningful decision for the wellbeing of the society. Given the aforementioned background, the book editors have selected an apt theme for the CBFS's fourth volume of an edited book, titled *Fourth Industrial Revolution and Business Dynamics: Issues and Implications*.

This book clearly demonstrates the viewpoints and ideas exceedingly well on some critical issues relating to 4IR. The book consists of fifteen chapters under four sub-themes that cover a large spectrum of contemporary topics. A wide variety of data and empirical information on issues and implications of Industry 4.0 presented in the book are highly relevant and useful to the present context. I certainly hope that the new body of knowledge introduced and developed by the authors in this book will enrich your knowledge and ideas about 4IR. I am sure that the book will also serve as an authoritative source of information and reference publication on 4IR for regulators, policy and decision makers, students and researchers. The initiative of publishing this book will certainly help the College of Banking and Financial Studies (CBFS) to develop a strong network among stakeholders.

I would like to place on the record my acknowledgement and appreciation to the College Management for taking up this knowledge transfer initiative for the benefit of the Community. This book is a good step in that direction. I also commend the considerable efforts of the editors and contributors for sharing their scholarly thoughts through this book. I wish the College the very best in all its research endeavours and look forward to many more contributions and scholarly publications from the College Community.

Dr. Khalfan Mohammed Al Barwani
Chairman of CBFS Board of Directors
Muscat, Sultanate of Oman

PREFACE

According to Schwab (2015), the First Industrial Revolution used water and steam power to mechanize production. The Second used electric power to create mass production. The Third used electronics and information technology to automate production. The founder and executive chairman of the World Economic Forum Prof. Klaus Schwab has coined the idea of Fourth Industrial Revolution which will be driven largely by the convergence of digital, biological and physical innovations. The Fourth Industrial Revolution's technologies, such as artificial intelligence, genome editing, augmented reality, robotics and 3-D printing, are rapidly changing the way humans create, exchange and distribute value. It is important to appreciate that the Fourth Industrial Revolution involves a systemic change across many sectors and aspects of human life. We are moving from automated production and IT, to an automated society. This revolution is expected to impact all disciplines, industries and economies. In this backdrop, we need to have a sound understanding Fourth Industrial Revolution and its economic, environmental and social impact and transforming strategies due to technological revolutions for survival and sustainability. This book seeks to bring together contributions in the form of chapters from researchers, academics and practitioners to provide useful input to develop future plans and strategies for meeting the challenges of the Fourth Industrial Revolution (4IR).

The book aims to explicitly explain strategic issues, trends, challenges and future scenarios of global economy in the light of 4IR. The book has been classified into four sub-themes, namely, digital innovation, sectoral impact, electronic business and employment and human resources. The book consists of fifteen chapters under the four sub-themes. These papers were selected after a double-blind peer review process. The chapters are authored by thirty-three scholars and practitioners across the globe. In this volume we have tried to cover a wide range of 4IR related issues such as big data, artificial intelligence, Robo-advisory, 3D printing, sectoral impact on healthcare, logistics, manufacturing, and education, adoption factor for E-Wallets, social media, risk of 4IR, reinventing human resources management, transformation in job market, human perspectives and talent management. We hope that the chapters of this book contribute to both academia and business by means of research, critical and theoretical reviews of issues in the Fourth Industrial Revolution. The chapters of the book will be indexed in leading research databases in due course and are available in Palgrave Macmillan, Springer Nature web portal.

The College of Banking and Financial Studies (CBFS), as part of its strategic goal, promotes research and encourages scholarly activities. Under its knowledge transfer initiative, CBFS successfully launched three volumes of the edited book on banking sector and financial sector and economic diversification in Oman in the past few years. This is the fourth volume published by CBFS under this series. These edited books have evoked positive response from both academia and industry. All these books are available for open access in CBFS portal and SSRN e-library.

We express our heartfelt gratitude to all the authors for sharing their invaluable thoughts by way of chapters in this edited book. We also extend our thanks to all the reviewers of the chapters of the book for their constructive feedback and to all those who have given their support in bringing this project to fruition. We also thank Ms. Sandeep Kaur, Associate Editor of Palgrave Macmillan for her continuous support and guidance in publishing this book. We certainly hope that the insightful ideas of this book will receive widespread acceptance.

Muscat, Oman

Nasser Rashad Al Mawali
Anis Moosa Al Lawati
Ananda S

CONTENTS

Editors and Contributors

About the Editors

Nasser Rashad Al Mawali is the former Dean of the College of Banking and Financial Studies, Muscat. He is a post doctorate from Oxford University (UK) on International Economics. He has a Ph.D. from Canberra University (Australia) on International Economics. He holds a master's degree in Economic Development from Vanderbilt University (USA) and a bachelor's degree in economics from Sultan Qaboos University (SQU). He was for twenty years a faculty member of Economics and was the Director of The Humanities Research Centre at SQU. He was a Visiting Scholar at Oxford University in 2014–2015. He was the Chairman of the Academic Group of the Indian Ocean Rim for Regional Cooperation for five years. He has participated in various government and non-government committees. Recently, he was appointed by the Council of Ministers as Board Member of The Competition Protection and Monopoly Prevention Centre. He has published numerous scientific refereed articles in international journals as well as a series of economic articles published in local newspapers and magazines and has participated in televised debates discussing local socio-economic issues. He has received a number of local and international awards, including the Best Economics Award for the year 2015 and the Best Paper Scientific Award at the IBERC Conference in the United States in 2000.

Anis Moosa Al Lawati is the Assistant Dean for Academic Affairs at the College of Banking and Financial Studies, Muscat. His B.A. is from Rollins College, Florida, USA (1986), and his M.B.A. from Yarmouk University, Jordan (1996). His areas of specialization include Management, Marketing, Operations Research and Human Resources. He has more than 30 years of experience in teaching and training in Oman, UAE and Jordan. He has previously held many key positions, including Deputy Dean of Oman College of Management and Technology, Director of Training and Learning at Dubai Institute of HRM, Acting Director General at IBFS, Director of Training at Institute of Public Administration and Acting Dean at CBFS. He conducts training in the areas of sales skills, marketing customer service, leadership, negotiation skills, supervisory skills and team management.

Ananda S is the Director of the Postgraduate Studies and Research Department at the College of Banking and Financial Studies, Muscat, Sultanate of Oman. He has a Ph.D. in Finance, with over thirty years of expertise in management education, consulting and the financial services industry. He has also been a visiting professor at leading B-schools in Dubai, Germany and India. He has published around 50 articles in peer reviewed journals indexed by *Scopus*, *Web of Sciences* and *ABDC* ranking list; and presented 40 research papers in national and international conferences. He has written books on Mutual Funds, Mergers & Acquisitions and Financial Systems & Commercial Banking and edited books on *Banking Sector of Oman, Financial Sector of Oman and Diversification of Oman Economy*. He is a Guest Editor, member of the Editorial Board and Reviewer for refereed *International Research Journals*. He has been a member of the subject expert committee to review the master and undergraduate curriculum in universities in India & Oman. He has also conducted training courses for leading corporates in India and Oman.

Contributors

Mawih Kareem Al Ani College of Commerce and Business Administration, Dhofar University, Salalah, Sultanate of Oman

Zainab Said Al Awaeed College of Commerce and Business Administration, Dhofar University, Salalah, Sultanate of Oman

Zainab Al Balushi College of Economics and Political Science, Sultan Qaboos University, Muscat, Sultanate of Oman

Anwar Al Sheyadi College of Applied Sciences, University of Technology and Applied Sciences, Rustaq, Sultanate of Oman

Ali Al Shidhani Ministry of Transportation and Communication and Information Technology, Muscat, Sultanate of Oman

Firas Almasri College of Arts and Sciences, Gulf University for Science and Technology, Kuwait City, Kuwait

Virginia Bodolica The Said T. Khoury Chair of Leadership Studies, School of Business Administration, American University of Sharjah, Sharjah, UAE

Elizabeth Chacko Center for Management Studies, Jain University, Bangalore, India

Gopalakrishnan Chinnasamy Department of Business & Accounting, Muscat College, Muscat, Oman

Neha Chitte Goa Business School, JRF Scholar, Goa University, Panaji, India

Gaitri Chugh Muscat College, Muscat, Oman

Tamanna Dalwai Muscat College, Muscat, Oman

Ashavaree Das Applied Media Division, Dubai Women's College, Higher Colleges of Technology, Dubai, United Arab Emirates

Syed Waheedullah Ghori Mechanical Engineering Department, College of Engineering, King Khalid University, Abha, Saudi Arabia

Shobhna Gupta Professional Studies and Undergraduate Department, College of Banking and Financial Studies, Muscat, Sultanate of Oman

Gertrude I. Hewapathirana College of Business Administration, Gulf University for Science and Technology, Kuwait City, Kuwait

Rabia Imran Department of Management, College of Commerce and Business Administration, Dhofar University, Salalah, Sultanate of Oman

Suaad Jassem Department of Managerial and Financial Sciences, Al-Zahra College for Women, Muscat, Sultanate of Oman

Mythili Kolluru Professional Studies and Undergraduate Department, College of Banking and Financial Studies, Muscat, Sultanate of Oman

Ashok Krishnan Central Queensland University, Rockhampton, QLD, Australia

Reshmy Krishnan Muscat College, Muscat, Sultanate of Oman

Shreesha Mairaru Applied Media Division, Dubai Women's College, Higher Colleges of Technology, Dubai, United Arab Emirates

Priya Makhija Center for Management Studies, Jain University, Bangalore, India

Syeeda Shafiya Mohammadi Muscat College, Muscat, Oman

Girija Narasimhan University of Technology and Applied Science, Muscat, Sultanate of Oman

R. Nirmala Goa Business School, Goa University, Panaji, India

Mohammad Rezaur Razzak Department of Management, College of Economics and Political Science, Sultan Qaboos University, Muscat, Sultanate of Oman

Preeti Shrivastava Department of Business & Accounting, Muscat College, Muscat, Oman

Nitha Siju Department of Business & Accounting, Muscat College, Muscat, Oman

Shreshtha Singhvi Infomerics Valuation and Rating Pvt. Ltd., Mumbai, India

Alan Somerville University of Stirling, Stirling, UK

Martin Spraggon School of Business and Quality Management, Hamdan Bin Mohammed Smart University, Dubai, UAE

Vineet Tirth Mechanical Engineering Department, College of Engineering, King Khalid University, Abha, Saudi Arabia

List of Figures

Big Data and Organizational Ambidexterity: A Strategic Perspective

Understanding the Emerging Role and Importance of Robo-advisory: A Case Study Approach

Healthcare Governance in the 4th Industrial Revolution: Leading Through Patient-Centeredness and Empowerment in the UAE

Cost–Benefit Analysis and Environmental Impact Assessment of 3D Printing Applications in Building Construction in Oman

Driving Factors of Adopting 4.0 IR Technologies in the Logistics Sector

Industry 4.0: The Future of Manufacturing—Foundational Technologies, Adoption Challenges, and Future Research Directions

Big Data Analytics and Accounting Education: A Systematic Literature Review

Diffusion and Adoption of E-wallets in Oman for Sustainable Growth

Present Practices and Future Challenges in Social Media Usage for Business: Observations from the United Arab Emirates

Re-Inventing Human Resource Management Through Artificial Intelligence

Industry 4.0: The Human Resource Perspective

LIST OF TABLES

Driving Factors of Adopting 4.0 IR Technologies in the Logistics Sector

The Prospects and Risks of Industry 4.0: Issues and Implications

Re-Inventing Human Resource Management Through Artificial Intelligence

Industrial Revolution 4.0: Transformation of Job Market

Talent Development Challenges and Opportunities in the 4th Industrial Revolution: A Boundaryless Career Theory Perspectives

Digital Innovation

Big Data and Organizational Ambidexterity: A Strategic Perspective

Mawih Kareem Al Ani, Rabia Imran,
and Zainab Said Al Awaeed

1 INTRODUCTION

Drawing upon the dynamics of the knowledge-driven economy the current research conceptualizes a strategic perspective whereby big data and organizational ambidexterity play their role. This chapter aims to explore the impact created by the presence of big data and ambidexterity on various outcomes of strategic nature. Moreover, it proposes a model depicting a holistic picture of organizational success. The worth of the proposed framework can be determined through its ability to exploit organization's full potential to achieve success. It is assumed that big data and organizational ambidexterity are involved in the whole strategic process and with the use of big data techniques along with organizational ambidexterity several benefits of strategic nature are achieved.

In this chapter, the big data is defined in terms of its five dimensions; velocity, value, volume, variety and veracity while, organizational

M. K. Al Ani (✉) · R. Imran · Z. S. Al Awaeed
College of Commerce and Business Administration, Dhofar University, Salalah, Sultanate of Oman
e-mail: mawih@du.edu.om

R. Imran
e-mail: rimran@du.edu.om

3

N. R. Al Mawali et al. (eds.), *Fourth Industrial Revolution and Business Dynamics*, https://doi.org/10.1007/978-981-16-3250-1_1

ambidexterity is defined as organization's ability to allocate the resources to exploit the existing opportunities and explore the new opportunities. A strategic perspective of big data and organizational ambidexterity will cover many issues such as firm successful, innovation, firm performance and competitive advantages

2   Importance and Definition of Big Data

In the current digital world data is available for almost every step taken by the people. The concept of big data is becoming the center of attention due to its unique nature. Big data is not just collecting of the information, but it is also concerned with its efficient usage (Krimpmann & Stühmeier, 2017). It is widely agreed that big data requires special handling. When the concepts of volume, variety and velocity are included then gathering information becomes more complex (McAfee & Brynjolfsson, 2012). It's special in nature due to being larger in size than the traditional databases so it requires specific techniques to handle (Watson, 2014). Unconventional softwares and systems are required because of the process of storing and processing of big data.

2.1   Historical Evolution of Big Data

The story of big data begins with information explosion that was the first attempted to measure the rate of growth in data volume. Currently, organizations are becoming increasingly aware of its importance as it may aid in capturing value of business and employees through processing and analyzing large amount of data (George et al., 2014). Table 1 shows the historical evolution of big data.

2.2   Dimensions of Big Data

Big data requires a difficult and complex process to analyze it. Sometimes the process is also not affordable, as there are many different sets of databases that need to be analyzed. Big data can be classified into the following dimensions called as 5Vs (Anuradha, 2015):

1. *Value*: This is one of big data's most important features; it is considered as its heartbeat. Big data is huge and it becomes useless, unless converted into some value. In order to create that value it is important to build IT systems and structures.

Table 1   Historical development of big data

Year	Event
1944	Ryder of Wellesleyan University has published a study on the future of research libraries in the United States
1967	The need to provide all the requirements for automatic and rapid storage has increased with the increase in the rate of transfer of information through the computer
1975	Japan's Ministry of Posts and Telecommunications managed the flow of information and tracked the volume of information in circulation
1978	The demand for information provided by the media and communication is characterized as being two-way
1981	The Central Statistics Office has launched a project to calculate the country's information industry
1993	Stefan Dinas has compiled a Unified System Manual for National Accounts
1996	Digital storage has become more cost-effective than paper storage
2000	Peter Lehmann and Halvarian published a study on how much information to determine the amount of computer storage
2005	• Promotes the concept of big data • Tim O'Reilly published an article entitled What is the second generation of the Web
2007	Researchers at IDC published a paper entitled Digital that is expanding globally: forecasts for information growth worldwide during 2010
2008	Cloud computing has been more innovative in its ability to collect, organize and process data in all areas of life
2009	Publication of a report on American consumers entitled Quantity of Information
2010	• Economist magazine published a special report entitled Data Everywhere • Scientists and computer engineers have created a new concept, big data
2012	The concept of big data has become such a technological and scientific cultural phenomenon
2013	Communicating things together is 7%, or 178 billion
Nowadays	Talk about the Internet of Things and its contribution to global data inflation has become

Source George et al. (2014), Krimpmann and Stühmeier (2017)

2. *Velocity*: As it is known, data flows at high speed. For example, within a minute, a huge amount of data flows. In this regard, social media is used within a very record time, to create files for this data, and put them in databases. It should be noted that speed is defined as the speed of data flow, as well as the speed of processing, and the

speed of decision-making in a very short time which is very difficult to process using conventional systems.

3. *Variety*: Data sources are highly heterogeneous, as they are arranged in different and diverse forms; they may be structured, semi-structured, or unstructured such as: written texts, recording, video and download files and others.

4. *Volume*: The volume of data increases day by day dramatically, reaching sizes MB, PB, YB, ZB, KB, TB and in an attempt to reduce the cost of storage data is produced in large sizes. Data is expected to increase 50-fold by 2020.

5. *Veracity*: When dealing with a large volume of data at high speed, the accuracy and reliability of the data will not be achieved by 100% and here the data becomes annoying and boring for the user of this data. Credibility can be achieved by pairing big data with data analysis techniques.

Figure 1 illustrates 5Vs of big data.

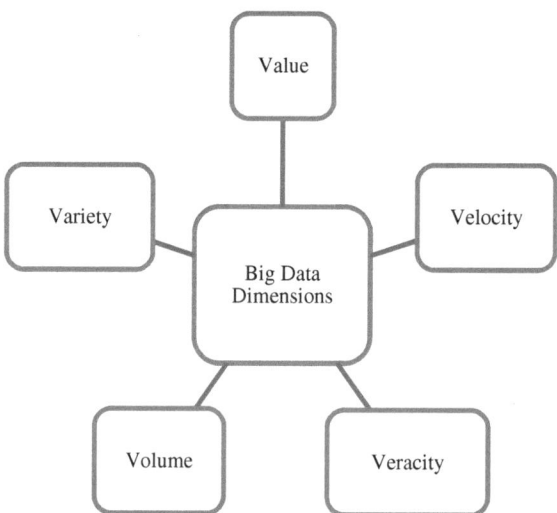

Fig. 1   Big data dimensions (*Source* Personal Collection)

3 Organizational Ambidexterity

This concept has been centering of attention recently. However, it existed for decades as a neglected concept. Recently the scholars started to argue about the factors of the long-term success of the organization and found organizational ambidexterity as one of the most important factors (Luo et al., 2017). Organizational ambidexterity can be defined in multiple ways. It can be described as an ability of an organization to allocate resources for successful exploitation and exploration to ensure its survival in the business environment (Yigit, 2013). It can also be defined as ability of the organization to explore and expand opportunities, to be transferred to a better reality and thus maximizing its role and importance in business environment (Katila & Ahuja, 2002). However, majority of literature defines it in terms of exploitation and exploration activities (Gupta et al., 2006; Jansen et al., 2005). Search for latest knowledge, potentials and prospects is part of exploration whereas, revision of existing capabilities and competencies is part of exploitation (Luo et al., 2017).

Figure 2 shows the components of organizational ambidexterity.

3.1 Exploration Activities

Exploration calls for search for new opportunities. It is the ability of being able to discover something and being able to find innovative ways of doing things (Yigit, 2013). Exploration activities are behaviors of the organizations characterized by search, risk taking and experimentation (Cheng & Van de Ven, 1996; March, 1991). In order to perform this type of activity successfully, the organization should firstly, mobilize

Fig. 2 Components of organizational ambidexterity (*Source* Personal Collection)

its resources and efforts for new opportunities (Dess et al., 2007) and secondly, establish the ability to identify and acquire external resources (Hsu et al., 2013).

3.2 Exploitation Activities

These activities call for the optimization of opportunities. Exploitation represents the ability of the organizations to improve activities of short-term value creation, designed for meeting existing customers' needs and seeks to expand current abilities along with expanding existing products and services, while increasing channels of expansion. The activities like effectiveness, production, selection, implementation and refinement are associated with exploitation (Cheng & Van de Ven, 1996). In this regard, the organization can invest some opportunities such as (a) having new customers, (b) expansion of the market through existing goods and services or through the introduction of new products (c) ability to diversify into goods and services (d) development of new technologies that can increase productivity and quality.

There are several steps to conduct the exploitation activities (Tabeau et al., 2017). These steps are as follows:

1. Identify the difficulties or obstacles that may hinder the exploitation of each opportunity and then identify the realistic opportunities that can be exploited in practice.
2. Consider a strategy through which the opportunity can be exploited through a focused growth strategy by increasing current sales in present markets, marketing of already existing services in some other markets or developing new market segments, developing existing services and providing new services in existing markets.
3. Objectives to exploit opportunities should be defined considering realism, quantifiable, time-bound, no conflict or overlap and ranked according to their relative importance.
4. Thinking about the strategy of exploiting opportunities and be organized in a time-bound manner with the development of policies and plans to do so.

The relationship between exploration and exploitation is very essential for the organization. These are independent dimensions requiring different structures. The firm might get some benefits from the previous investments in exploration process to maximize these benefits in the future. For example, if the firm has one project, it will try to avoid all previous shortcomings to maximize the value for the firm (Popadić et al., 2015). The emphasis of exploitation activities is on stability, receptive customers, need of efficiency and process reliability. However, exploration activities are focused on coping with environmental changes, innovations and varying requirements. Basically, exploitative activities are based on available knowledge with an attention toward enhancing the already available expertise, practice, structures, customers or markets. For example, if current customer is the main focus of the firm, the exploitative activities should design to meet current customers' needs. On the other side, exploratory activities are undertaken to create new products and develop new designs to meet emerging customers or market's needs (Benner & Tushman, 2003). Both the activities bring benefits for the organizations in the long term.

The available literature discussed the trade-off between exploitation and exploration. The balancing between those two types of activities is in the core of organizational ambidexterity. The organizations that are not able to maintain this trade-off face a downward spiral as its performance as its survival is based on achieving a balance between exploration and exploitation activities (Hughes, 2018). Both exploration and exploitation activities have a different perspective and relationship with other organizational variables and lead to different results (Gupta et al., 2006). Table 2 summarizes the differences between the exploration and exploitation activities:

4   Big Data and Organizational Ambidexterity

Big data and organizational ambidexterity results are two important variables whose presence in the organizational environment can do wonders. This combination is a holistic approach that can result in improved organizational processes and organizational success. This is a unique combination that can affect multiple outcomes and result in sustainable organizations. In this part of the chapter, we will discuss some of them.

Table 2 Difference between exploration and exploitation activities

Exploration activities	Exploitation activities	Statement	
Activities designed to meet the needs of customers and new markets	Activities designed to meet the needs of existing customers and markets	Definition	1
New designs, new markets and new distribution channels	Existing designs, existing markets and distribution channels	Outputs	2
Require new knowledge and gain from existing knowledge	Adopt and expand current knowledge and skills	Knowledge Base	3
Research, flexibility, scientific experiments and risk	Refinement, production, efficiency and implementation	Produced from	4
Long-term	Short-term	Performance applications	5

Source Gupta et al. (2006), Hughes (2018)

4.1 Big Data, Organizational Ambidexterity and Firm Success

One of the long-term objectives of a firm is its success. Basically, firm success means that the firm has the ability to achieve its objectives within its strategic framework which may be difficult due to a number of reasons; first, review of a lot of information is required to avoid mistakes in decision-making process. Second, a lot of resources as inputs are required for this process. Third, due to the dynamic environment, it becomes difficult to develop exactly predictive forecasting models. Due to all these reasons, organizations would prefer to allocate resources for exploitation and exploration which are the dimensions of ambidexterity. Most of the studies in the area of ambidexterity have decided that the ambidexterity is associated positively with longer survival and firm success (Cottrell & Nault, 2004). Thus, although ambidexterity is one of the most important difficulties and it is a challenge to execute it in the appropriate strategic context, for sustained competitive advantages.

But what is the role of big data in this regard? Globally, a big number of firms that are using big data techniques invested a lot of resources in these techniques. Big data techniques change the information generation process which would aid in futuristic decision-making by the organizations. Futuristic decision-making requires big data (volume and variety) to analyze different scenarios including but not limited to new products and

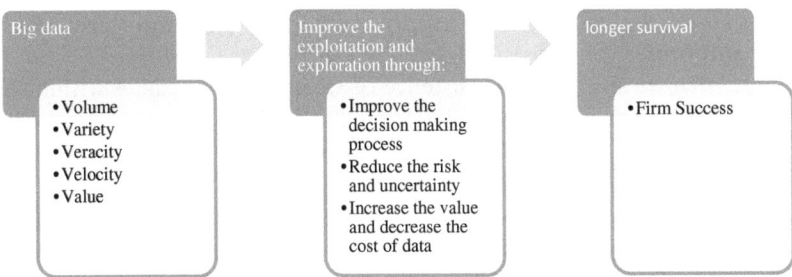

Fig. 3 The relationship between big data, organizational ambidexterity and firm success (*Source* Personal Collection)

marketplace analysis and development of the existing projects (Jebel et al., 2018). The use of big data also reduces uncertainty and risk and ensures better decision-making. Popovič et al. (2018) pointed out that "within the current turbulent and highly competitive global environment, firms are compelled to adapt more rapidly, boldly, and to experiment in order to survive and thrive." In this regard, the firms seek to collect process and evaluate the data as fast as possible within the framework of data velocity. Finally, for any decision of the future of the firms, the cost of the data should be less than its value which means that the firms have an ability to maximize the benefits of the data. This will allow the firms to improve their opportunities to exploit and explore the new projects and existing projects. Figure 3 conceptualizes the relationship between big data, organizational ambidexterity and firm success.

4.2 Big Data, Organizational Ambidexterity and Innovation

Innovation is vital for the success of an organization. It is the key to its sustainability. Organizations need to innovate due to a number of reasons. First, it will improve the competitive position of the firm by introduction of new ideas, services and products. This will enhance the relationship with the customers and market and finally beat the competitors. Second, it will give the firms sustainability through the continual improvements of the process and operations. In total, Markides and Chu (2009) find out that the innovation supports and promotes the organizational ambidexterity through the following:

a. *Autonomy solutions*: when the center of the firm gives more authorities to its divisions.
b. *Cultural solution*: the existence of strong, shared values within the organization would allow corporate headquarters to grant autonomy to divisions without losing control over them.
c. *Communication solution*: the frequent communication, frequent rotation of managers and corporate-sponsored training programs could all be employed as integrative mechanisms to improve the decision-making process.

The relationship between Innovation and ambidexterity is the key strategic component (especially for the firms that have an international perspective as these firms have greater customer base with multiple demographic profiles. This relationship has been explored by a few research (Scott, 2014; Stettner & Lavie, 2014; Yu et al., 2014). However, very few researchers have identified the presence of big data and organizational ambidexterity in relation to innovation. For example, Bøe-Lillegraven (2014) explains the relationship between Innovation, ambidexterity and big data by using one of big data characteristics which is velocity. He pointed out that the high-velocity data will allow for the continuous analysis of the micro-foundations of explorative activities. For example, the firm can use a flexible budget to allow finding new ideas and new products and this will help the inventor to use the resources in developing these ideas and products. The flexible budget has a positive relationship with velocity data where the firm can change the scenario as it wants. The big data could track the current scenario to find new scenario which is suitable for the benefit of the firm. Another explanation of this relationship was introduced by Popadić et al. (2015). They pointed out that the firms can use big data features to improve the relationship between ambidexterity and innovation performance. For example, the exploitative activities aimed at improving existing product-market domains and exploratory innovation as technological innovation aimed at entering new product-market domains. Figure 4 conceptualizes the relationship between Innovation, ambidexterity and big data.

Fig. 4 The relationship between ambidexterity, innovation and big data (*Source* Personal Collection)

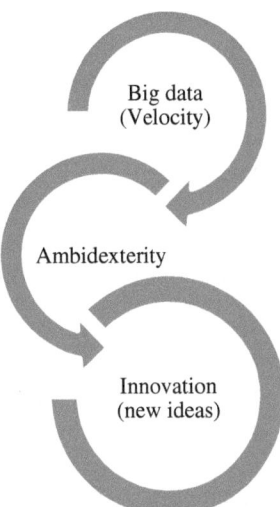

4.3 Big Data, Organizational Ambidexterity and Performance

Most of the studies in the area of ambidextrous and performance confirm that an ambidextrous strategy has a positive effect on organizational performance. Peng et al. (2019) supported this idea when an organization gets involved in exploration and exploitation actions then its performance gets better because exploration is the basis of organizational growth. Firms in a competitive environment having limited resources but industrial development will devote themselves to seek opportunities for development, growth and promoting innovation. This will help the firms not only to improve their operational efficiency and effectiveness but also promotes innovative performance (environmental adjustment, new market development, new product). Gupta et al. (2006) pointed out that the firm performance will not improve without increasing the ability of the firm to "engage in enough exploitation to ensure the firm's current viability and to engage in enough exploration to ensure future viability." Tokgöz et al. (2017) introduced some evidence about the relationship between ambidextrous and some common types of performance such as financial performance and marketing performance. For example, the exploration activities will improve the marketing performance by developing new possibilities that goes beyond the current situation of markets, products, technologies and capabilities.

The relationship between ambidexterity and performance does not exist in isolation. Presence of big data has a big role to play here. There are many areas that big data can improve the firm performance:

1. The big data improves the accuracy of forecasting of sales in terms number of products and time of offer the products for sales (Bajari et al., 2019).
2. The big data analytics can increase the effectiveness and efficiency of firms.
3. The big data enhances the customer–relationship management.
4. The big data can moderate the operational costs and improve quality of life.
5. The big data improves supply-chain management.
6. The big data analytics can optimize prices; increase profit and maximize sales, financial productivity and market share as well as return on investment (Maroufkhani et al., 2019).

Positively, the firms that used big data analytics were very effective in terms of organizational ambidextrous by developing the exploration and exploitation activities and the result was that the firm performance is improved. In order to explain the relationship between organizational ambidextrous, big data and performance, we will give one explanation by using one of big data features which is the value of data. The value of big data is significant since the cost–benefit criterion is one of the most important criteria to enhance firm performance. Basically, the benefit of the data should be greater than its cost then the decision-maker will use these data to build appropriate data models to maximize the performance of the firm. In those models, the ambidextrous needs a high-value data to take a decision about exploration and exploitation activities such as introduce new products, new services and new technologies then the firm will calculate the return from each activity to select the activity which maximizes the performance. Figure 5 conceptualizes the relationship between ambidexterity, performance and big data.

Fig. 5 Relationship
between ambidexterity,
performance and big
data (*Source* Personal
Collection)

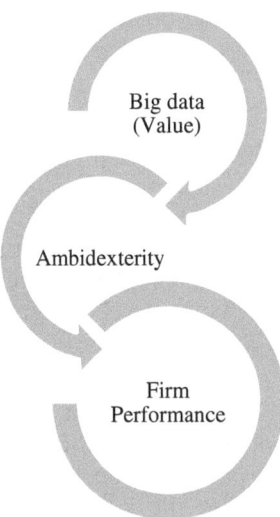

Big data
(Value)

Ambidexterity

Firm
Performance

4.4   *Big Data, Organizational Ambidexterity and Competitive Advantage*

A competitive advantage exists when the firm can deliver the same benefits as competitors with a different feature from customer's perspective. For example, the firm can deliver the product to the customer at lower level of cost (cost advantage) (Wen-Cheng et al., 2011). Preda (2014) pointed out that many of the firms across the world tried to develop new ideas, new products and enter to the new market. This issue has a priority for the firms and most of these firms worked hard to achieve their objectives and gain competitive advantage. The concept of organizational ambidexterity is one of the dynamic capabilities of the firm and can be found as a combination of two different activities: exploration and exploitation. There are some areas where the organizational ambidexterity will facilitate the obtaining of competitive advantage (Preda, 2014):

1. The organizational ambidexterity will bring many innovative ideas through exploration and exploitation activities.
2. The organizational ambidexterity will introduce many new products through exploration and exploitation activities.

3. The organizational ambidexterity will enable the firm to enter the new markets through exploration and exploitation activities.

There are many results concluded from this relationship between organizational ambidexterity and competitive advantages:

1. High level of profit, market share value and sales growth
2. Increase the market share of the firm
3. Improvement of the customer relationships
4. Lower level of costs
5. Improvement of firm performance.

The role of big data in the relationship between organizational ambidexterity and competitive advantages is very important. One of the possible probabilities of this relationship is the role of big data in building the knowledge assets in the firm. Kamioka and Tapanainen (2014) pointed out that many organizations used big data to build knowledge infrastructure and all other knowledge activities to influence the competitive advantages. This issue needs high-speed data, high amount of data and high quality of data where the firms can build the knowledge and then acquire competitive advantages. Yadav and Pavlou (2014) pointed out the possibilities of Big Data for marketing activities. They discussed that by providing information about customers, products and markets, the firm will gain competitive advantages. If the firm has full details and statistics about these three things, the decision-makers will be able to build the right decision model to the favor of business. Figure 6 conceptualizes the relationship between ambidexterity, competitive advantages and big data.

5   Conclusion

In this chapter, the relationship between big data and organizational ambidexterity was discussed. Moreover, the relationship between some strategic objectives such as firm's success, firm performance, innovation and competitive advantage was discussed. It is evident that if the organization is interested in its long-term survival then it must engage in organizational ambidexterity. The use of big data then becomes a necessary part of its decision-making process. In this way, organizations become more flexible to have a plan with very important strategic objectives.

Fig. 6 Relationship between ambidexterity, competitive advantages and big data (*Source* Personal Collection)

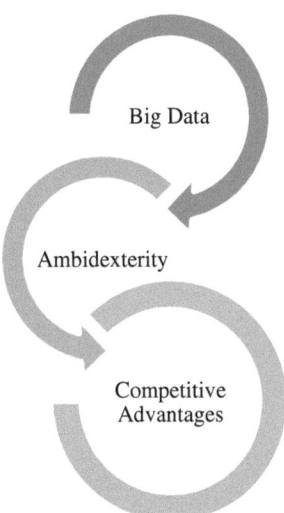

Many of these objectives in order to be achieved need a lot of information. The organizational ambidexterity is defined in terms of exploration and exploitation activities while the big data is defined in terms of five dimensions namely, value, velocity, variety, volume and veracity. The presence of these two variables can create wonders for the organizations.

In the final part of the chapter, the relationship between the big data, organizational ambidexterity and the strategic objectives was discussed separately. The most important result is that by using big data and organizational ambidexterity the organization will be able to achieve the innovation, competitive advantage, high level of performance, and finally organizational success. Thus, this chapter proposes a conceptual model that explains that the presence of both Big data analytics and organizational ambidexterity creates a positive influence on innovation which result in competitive advantage which in turn increase firm performance and result in long-term firm success. This model as shown in Fig. 7 can be empirically tested in future research.

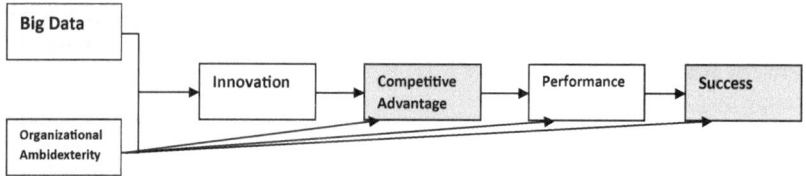

Fig. 7 The proposed holistic model (*Source* Personal Collection)

References

Anuradha, J. (2015). A brief introduction on Big data 5Vs characteristics and Hadoop technology. *Procedia Computer Science, 48*(Special Issue), 319–324.

Bajari, P., Victor, C., Ali, H., & Junichi, S. (2019). The impact of Big data on firm performance: An empirical investigation. *AEA Papers and Proceedings, 109,* 33–37. https://doi.org/10.1257/pandp.20191000.

Benner, M. J., & Tushman, M. L. (2003). Exploitation, exploration and process management: The productivity dilemma revisited. *Academy of Management Review, 28*(2), 238–256.

Bøe-Lillegraven, T. (2014). Untangling the ambidexterity dilemma through big data analytics. *Journal of Organization Design, 3*(3), 27–37. https://doi.org/10.7146/jod.18173.

Cheng, Y. T., & Van de Ven, A. H. (1996). Learning the innovation journey: Order out of chaos? *Organization Science, 7*(6), 593–614.

Cottrell, T., & Nault, B. R. (2004). Product variety and firm survival in the microcomputer software industry. *Strategic Management Journal, 25*(10), 1005–1025.

Dess, G., Lumpkin, G. T., Eisner, A. B., & McNamara, G. (2007). *Strategic management: Text and cases*. McGraw-Hill Irwin.

George, G., Haas, M. R., & Pentaland, A. (2014). Big Data and management: From the editors. *Academy of Management Journal, 57*(2), 321–326.

Gupta, A. K., Smith, K. G., & Shalley, C. E. (2006). The interplay between exploration and exploitation. *Academy of Management Journal, 49*(4), 693–706.

Hsu, C. W., Lien, Y. C., & Chen, H. (2013). International ambidexterity and firm performance in small emerging economies. *Journal of World Business, 48*(1), 170–176.

Hughes, M. (2018). Organizational ambidexterity and firm performance: Burning research questions for marketing scholars. *Journal of Marketing Management, 34*(1–2), 178–229. https://doi.org/10.1080/0267257X.2018.1441175.

Jansen, J. J., Van den Bosch, F. A., & Volberda, H. W. (2005). Exploratory innovation, exploitative innovation, and ambidexterity: The impact of environmental and organizational antecedents. *Schmalenbach Business Review, 57*(4), 351–363.

Jeble, S., Kumari, S., & Patil, Y. (2018). Role of big data in decision making. *Operations and Supply Chain Management, 11*(1), 36–44.

Kamioka, T., & Tapanainen, T. (2014). Organizational use of big data and competitive advantage-exploration of antecedents. In *PACIS 2014 Proceedings*, 372. http://aisel.aisnet.org/pacis2014/372.

Katila, R., & Ahuja, G. (2002). Something old, something new: A longitudinal study of search behavior and new product introduction. *Academy of Management Journal, 45*(6), 1183–1194.

Krimpmann, D., & Stühmeier, A. (2017). Big data and analytics: Why an IT organization requires dedicated roles to drive sustainable competitive advantage. *International Journal of Service Science, Management, Engineering, and Technology (IJSSMET), 8*(3), 79–92.

Luo, C., Zhang, D., Luo, B., & Ge, J. (2017). Ambidextrous strategy and firm performance: The moderating effects of organizational slack and organizational life cycle. *Business and Management Studies, 3*(4), 1–11.

March, J. G. (1991). Exploration and exploitation in organizational learning. *Organization Science, 2*(1), 71–87.

Markides, C., & Chu, W. (2009). Innovation through ambidexterity: How to achieve the ambidextrous organization. In L. Costanzo & B. MacKay (Eds.), *Handbook of research on strategy and foresight*. Elgar.

Maroufkhani, P., Wagner, R., Ismail, W. K. W., Mas, B. B., & Nourani, M. (2019). Big data analytics and firm performance: A systematic review. *Information, 10*(7), 226. https://doi.org/10.3390/info10070226.

McAfee, A., & Brynjolfsson, E. (2012). Big data: The management revolution. *Harvard Business Review, 90*(10), 60–68.

Peng, M.Y.-P., Lin, K.-H., Peng, D. L., & Chen, P. (2019). Linking organizational ambidexterity and performance: The drivers of sustainability in high-tech firms. *Sustainability, 11*, 3931. https://doi.org/10.3390/su11143931.

Popadić, M., Černe, M., & Milohnić, I. (2015). Organizational ambidexterity, exploration, exploitation and firms innovation performance. *Organizacija, 48*(2), 112–119. https://doi.org/10.1515/orga-2015-0006.

Popovič, A., Hackney, R., Tassabehji, R., & Castelli, M. (2018). The impact of big data analytics on firms' high value business performance. *Information Systems Frontiers, 20*(2), 209–222. https://doi.org/10.1007/s10796-016-9720-4.

Preda, G. (2014). Organizational ambidexterity and competitive advantage: Toward a research model. *Management & Marketing, XI, I*(1), 67–74.

Scott, N. (2014). Ambidextrous strategies and innovation priorities: Adequately priming the pump for continual innovation. *Technology Innovation Management Review, 4*(7), 44–51.

Stettner, U., & Lavie, D. (2014). Ambidexterity under scrutiny: Exploration and exploitation via internal organization, alliances, and acquisitions. *Strategic Management Journal, 35*(13), 1903–1929.

Tabeau, K., Gemser, G., Hultink, E. J., & Wijnberg, N. M. (2017). Exploration and exploitation activities for design innovation. *Journal of Marketing Management, 33*(3–4), 203–225. https://doi.org/10.1080/026 7257X.2016.1195855.

Tokgöz, E., Akatay, A., & Özdemir, S. (2017). Impact of ambidextrous marketing on market and financial performance. In *EBEEC Conference Proceedings, The Economies of Balkan and Eastern Europe Countries in the Changed World, KnE Social Sciences* (pp. 59–72). https://doi.org/10. 18502/kss.v1i2.647.

Watson, H. J. (2014). Tutorial: Big data analytics: Concepts, technologies, and applications. *Communications of the Association for Information Systems, 34*(1), 65.

Wen-Cheng, W., Chien-Hung, L., & Ying-Chien, C. (2011). Types of competitive advantage and analysis. *International Journal of Business and Management, 6*(5), 100–104. https://doi.org/10.5539/ijbm.v6n5p100.

Yadav, M. S., & Pavlou, P. A. (2014). Marketing in computer-mediated environments: Research synthesis and new directions. *Journal of Marketing, 78*(1), 20–40.

Yigit, M. (2013). *Organizational ambidexterity: Balancing exploitation and exploration in organizations.* Unpublished Master thesis. Entrepreneurship and Business Development, School of Management, Blekinge Institute of Technology, Sweden.

Yu, X., Chen, Y., Nguyen, B., & Zhang, W. (2014). Ties with government, strategic capability, and organizational ambidexterity: Evidence from China's information communication technology industry. *Information Technology and Management, 15*(2), 81–98.

Efficiency and Advancement of Artificial Intelligence in Service Sector with Special Reference to Banking Industry

Priya Makhija and Elizabeth Chacko

1   INTRODUCTION

In any economy to maintain sustainability, banking plays a pivotal role by maintaining citizenry's investment, wealth sector and trade growth as well as commercial banks have always been a productive area towards development in artificial intelligence and machine learning (Shu & Strassmann, 2005). The recent digital evolution undoubtedly has affected conventional banking systems, and at the same time, it has made companies to be insecure and susceptible to cybersecurity risks. In the development of an integrated financial fraud defence system companies are increasingly taking account of technological innovation such as blockchain (Jewandah, 2019). FinTech Trends India Survey by PwC shows that global investment in artificial intelligence had crossed 5.1 million dollars in the year 2017, it was increased to 41.1 billion dollars in 2018, and it is expected

P. Makhija (✉) · E. Chacko
Center for Management Studies, Jain University, Bangalore, India
e-mail: priya_m@cms.ac.in

E. Chacko
e-mail: elizabeth_c@cms.ac.in

N. R. Al Mawali et al. (eds.), *Fourth Industrial Revolution and Business Dynamics*, https://doi.org/10.1007/978-981-16-3250-1_2

21

to hit 300 billion dollars by 2030. United Services Automotive Association (USAA) is a first bank to incorporate clients' smartphone deposits, and also has agreed to invest in AI which will help to detect potential anomalies through apps (Walch, 2020) connected to fraudulent activities and fraud prevention (Padhi, 2019). AI drew quite a lot of investment, particularly in the US and China. In Europe, the view is still quite uneven, with some of those countries having a healthy AI environment, whereas others fall back. European governments have put in place initiatives to raise AI development in Europe.

According to Accenture IT may allow banks to cut costs to 25% by applying AI into business operations (Harman, 2019). Corporations have already shown that certain activities and maybe even their business strategies emerge through the use of AI (Bughin et al., 2018). AI has the ability to dramatically alter facets of investment banking. In such an era wherein innovation in banks are incredibly competitive—thanks to data driven financial institutions such as financial technology (FinTech) start-ups and massive tech giants confronting conventional banking business strategies where the successful evolution of Artificial intelligence could be crucial for banks to stay viable (Kaya, 2019). Banking AI is currently primarily seen in a small range of back-end operations, like stock forecasting and credit scores by major commercial banks or creditworthiness by credit card issuers which are probably highly digital (Jadhav et al., 2016).

Big Data analytical approaches are even used to assess the risk in collateral acceptance (Hwang et al., 2004) and criminal identification (Ngai et al., 2010). Bank financial crises can lead to economic contractions whereas efficacy of bank financing can bring financial stability as per researchers. As per Ataullah and Le (2006) an improvement in performance of banks globally was seen after economic reforms i.e. fiscal reform, financial reform and private investment liberalisation and A positive relationship was found between the degree of competition and the effectiveness of banks. Services originally given only through banks like processing payments (PayPal), currency conversion (Transferwise) and lending (MyBank) are now being carried out by non-banking institutions where businesses use artificial intelligence to help their services most reliable, competitive and in affordable way (Zhang & Woo, 2018) to meet the profitable goals (Dietz et al., 2016).

The transition and competitiveness have contributed to a weakened engagement between banks and customers, particularly millennials, small and medium-sized enterprises and individuals are lacking exposure to the banking sector (Dietz et al., 2016). Besides bankruptcy prediction, the amount and range of data styles augmented through AI and transaction data have a significant effect on firm performance of predictive analytics in the sense of computational defaults debt (Kvamme et al., 2018; Tobback & Martens, 2017). Banks can take this as an opportunity through AI banks should evaluate the financial actions of customers and offer their services appropriately in order to retain consumers and gain their trust (Martens et al., 2016).

AI Non-financial gains could also be established, such as the reduction in the workload of human staff by enhancing money laundering detectors, which could boost consumer safety and confidence and minimise the harm caused through laundering money, scandals (G. Kumar et al., 2019). It is projected that Artificial intelligence will produce up to 1 trillion dollars of economic value per year for international banking (Biswas et al., 2020), Although many banks are having trouble moving from experiments to scaling AI technology throughout the business and its main reasons are absence of a consistent AI Policy, a rigid and investment greedy tech centre, scattered client data and outdated business model that hamper business-technology cooperation (Chui & Malhotra, 2018). Disruptive AI innovations can greatly enhance the capacity of banks to accomplish four main results: higher profitability, customization in a range, Omni channel approach and intense levels of development (Biswas et al., 2020). On the contrary, as per McKinsey's Global AI Report 60 Per cent of commercial banking industry participants indicated their organisations have minimum one AI capability. The most popular AI applications are mostly automation and robotics (36%) for organised operational functions; robotic helpers or interaction frameworks (32%) for client service sections; and machine-learning techniques (25%) (Cam et al., 2019).

2   RESEARCH METHODOLOGY

This conceptual paper based on secondary sources tries to focus on the efficiency and advancement of Artificial Intelligence techniques which are used in Banking Industry. Through the research of the previous paper the following research gap has been incorporated as the objective of this paper:

(i) Growth of Artificial Intelligence tools or techniques in the Banking Sector and its impact on human resources.
(ii) Problems faced by banks to adopt the technologies like government regulations and encountering frauds.
(iii) Customer expectation from banks through adoption of AI techniques.

3   Literature Review

Artificial Intelligence commonly known as AI evolution came into existence in 1956 (Uzialko, 2019). In 1943 recognition of work regarding AI was by Warren McCulloch and Walter Pitts with the model named artificial neural. Slowly growth in technology built first neural computer in 1950s (Ferreira et al., 2020; Russell & Norvig, 2013; Turing, 1950). There were more technological formulations made during 80s and 90s (Ferreira et al., 2020), which made machine think and work like humans (Turing, 1950). *"Application of AI is endless"* (Frankenfield, 2020) as now every business has lifted towards the change and adaption of AIs in their industries (Khurshid, 2020). Manufacturing and production units introduced AI in their operations long back, that reduced human error and more cost effective. It can be healthcare sector, self-driven cars, hospitality, logistic, education, financial services, retail or customer care (Ferreira et al., 2020). With the growth of AIs there are more job avenues too opened up in customer service business (A. Kumar, 2017).

Chatbots and other features like face recognitions can help the clients and customers with fast responses and confidentiality. Financial sector, being the backbone of the economy has also updated to newer technologies. Banking sector has applied AIs in various operating areas (Bhushan, 2018; Ferreira et al., 2020) due to tech savvy customers, that bring the banks into virtual mode (Kaur et al., 2020). Customers these days are more relied on the technology and gadgets for day-to-day work. Bank adopted many technologies to provide customers hurdle free experiences (Khurshid, 2020). According to Wipro's report many financial service providers are using predictive analysis and voice recognition to harness the technologies in competition and digitization (Khurshid, 2020). Inclusion of AIs in bank has reduced the fraudulent activities Banks as AI is used in all the three channels of banks like front office for conversational banking, middle office for anti-fraud and back office for credit under-writing (Digalaki, 2019; Latimore, 2018).

AI application in banks have brought better relations with customer as it can manage languages barriers and resolve or report issues instantly, also follow up (Bhushan, 2018; Digalaki, 2019; Kaur et al., 2020). As per Latimore (2018) report inclusion of AI in bank can mitigate risk by error reduction (typo error, misreading), increase revenue, effective work by reducing cost, use of technology for analytics through robotic process automation (RPA) and chatbots. Banks use AI for predictive analysis for detecting fraud to prevent money laundering, also develop customers risk profiles and fraud scoring (Mejia, 2020). AI can also detect fraud in online forms applying for loans also report contextual fraud by categorising the type of fraud (Mejia, 2020). Various other frauds can be identity theft, money laundering, tax evasion, forgery, embezzlement etc. (Nandi, 2020). Adding to, PwC report on Fin tech trends in 2020, it is said that 80% of banking CEOs are apprehensive about the pace at which the technology changes (Garvey & Burns, 2020) than any other sector. The challenge for banks due to technological disruption will be to accept smart changes and customer preferences (robo investing, settlement of cash, change of bank).

Banks must operate on basis of sharing and collaboration to future growth (Garvey & Burns, 2020). As per PwC retail banking survey 2020, it is says only 17% of banks are prepared, 61% of bank executives say that any technology adapted must be customer centric, and globally 75% of banks are investing on AIs for future growth. The innovation in AI and ML will bring a significant change, also global purview of fraud detectors is increasing as in 2016 only 14.37 bn dollar, 2020 its 28.8 bn dollar and expected in 2022 will be 41.59 bn dollar (Kharod, 2020). This will improve the efficiency of banks and provide faster services to clients and customers.

As new inventions and innovations are taking birth, simultaneously the cybercrimes are also increasing. As per PwC global economic crime and fraud survey 2020 (Rivera et al., 2020) in which they interviewed 5000 respondents from 99 territories, it says 47% of the companies have experienced fraudulent actions. So banks need to upgrade and keep themselves up to date with the appropriate (Nandi, 2020; Rivera et al., 2020). The various other AI tools used to detect fraud is through bank details, unusual bank balance, spending records, that can be mapped (Nandi, 2020).

AI can also verify the identity issued by government and matching it with facial biometrics or photo or video, so that more authenticity is there from customers end and cyber security for the banks (Ibbitson, 2020). Furthermore (Maheshwari, 2020) says future banking will be combination of AI and HI (human intelligence) for better customer experience and to build trust. Inclination of banks towards AI can provide services to customers anywhere and anytime (24/7) (Gupta, 2020). Customer interaction through AI (voice banking, Chatbots, has enhanced the banking efficiency (conversational banking), reduction in unsettled complaints and able to study customers behaviour pattern based on voice patterns and past calls (Barthet, 2019; Chalimov, 2019; Gujral, 2020).

The Chatbots improved versions are adopted by banks these days as earlier ones were like pre-recorded data, that doesn't think, but these days the Chatbots converse with the customers like human's and are able to advise according to their problems (Gujral, 2020). AIs can understand the emotions of customer like if "very angry" can predict that escalation of the case is recommendable to avoid worse situations, this focusses more on personalised experiences that grows the relationship between bank and customer (Gujral, 2020; Gupta, 2020; Ibbitson, 2020). "Erica" a chatbot developed by Bank of America recognise speech and read text messages and provides financial advises (Chalimov, 2019). Many banks have applied "voice biometrics" like "Lloyds Bank, the Bank of Scotland, or Halifax UK" (Michałowska, 2020) as customers are shifting from physical visit to M-banking (mobile banking).

It is predicted that in the year 2023 over 7.33 billion people worldwide will be using smartphone, that also assumes that increase in revenue from mobile market might be 935.2 billion (Michałowska, 2020). Advancement in AIs and implication in financial sectors to one extent has brought seamless operations for customers and clients, but the rate at which cybercrime or fraud is into high peak. Frauds result to huge loss to banks, around 270,000 cases were reported against credit card fraud in 2019 (Trehan, 2020). Loan fraud is increasing as many bank applications for credit card or advance loans requires few personal information's, that can be forged.

Fraudulent mortgages are riskier for banks as per study 0.81% of the application have fraudulent information (Trehan, 2020; Zhou, 2020). To reduce frauds banks and technology companies are working on security issues. Capegemini suggested that proper inclusion of AI in operation can be reduce fraud investigating time by 70% (Trehan, 2020). Adding on google is applying advanced machine learning to reduce transactional frauds by alerting users regarding spam mails and phishing mails (Trehan,

2020). Banking frauds have risen upto 73.8% after all the efforts taken by government (DR, 2020a). In that also state owned banks had more fraudulent cases 148,400 crores in 2019–20, whereas the private banks fraud cases were 34,211 crores in 2019–20, comparing above two foreign banks has very less fraud attacks 972 crores in 2019–20 (PTI, 2019; Palepu 2020). Frauds under various category can be lending operations constitute to 98% fraud cases, cards and internet frauds have risen to 174%, deposit related frauds have raised to 316% (Palepu, 2020). RBI has to re-work on the implementation of fraud detecting machines for early warning signals (EWS) (Palepu, 2020). To fight against these frauds banks have to come together and share knowledge regarding AIs to foster banks operations and reduce losses (DR, 2020b) as majority of the frauds happen due to inadequate internal processes (PTI, 2019).

As per the comparison shown between foreign banks, state owned banks and private banks, foreign banks must share the knowledge regarding their operations to others for financial growth of the economy (DR, 2020b). The facts regarding fraud stated above might be cases filed, but there are times cases are recorded, as awareness regarding fraud incidents must be reported to Digital Wrongdoing branch (Darshnik, 2019). Many suggested a cluster of old rule book and layers of technology can avoid fraudulent activities like multi-factor authentication and giving awareness on fraud to clients and customers (McGlasson, 2010). Sensitive systems and anti-fraud systems will effective and efficient for clients and banks (Guardians, 2019). To sum up the future of banks and other industrial sectors depends upon AI's, for that technology need to be keep tracing the errors and find solutions for issues and frauds. The business focussing on implementing AI can survive rest will diminish.

4   Discussions

4.1   Growth of Artificial Intelligence

Reliable AI forecasts can be made using specially crafted data forms in the case of loans. By examining these overlooked data forms, risk management systems can become more reliable revenues can be improved and new types of customers can be served. Holistic proposals to focus on "jobs to be done" have to be organised by the banks (Christensen et al., 2016) which can be achieved by incorporating individualization choices, creating product offerings, Non-Banking products and services (Biswas et al.,

2020). There are now 24-hour AI chatbots and voice-enabled (Stephanie, 2019) enabled tools for any task which requires no interpersonal contact such as getting acquainted with banking services, solving issues and helping solutions with any issues they might have. Virtual Assistants learn much about commercial banks continuously by monitoring their past dealings and browsing habits to give highly customised user interfaces. The Erica launched by Bank of America Marous (2020) provides details about the path to saving money, sends reminders to your customers, alerts consumers about changes to credit reports, lets them pay bills and instantly transfers. Erica actively learns and delivers appropriate content and tailored customer service based on new materials about their stakeholders' requirements. SBI is presently operating Chapdex SIA (Ghosh, 2017), an Intelligence application which answers consumer demands immediately and supports clients with everyday banking tasks. EVA (Electronic Virtual Assistant) (Parayil, 2018; Sharma, 2018) chatbot created by HDFC Bank has been answering over 2.7 million customer requests, communicating with more than 530,000 unique users and holding 1.2 million discussions. iPal started by the ICICI Bank (Maru, 2017). For about 6 million requests, the chatbot communicated with 3.1 million subscribers at 90% accuracy. In order to assist clients with financial and non-Financial transactions, Axis Bank introduced an AI & NLP (Natural Language Processing). Software, Contextual Financing. Receiving the creative AI innovative technologies for the Banking industry, the Axis Bank, India's third largest private sector bank launched last year an Innovation lab called Think Factory (Baruah, 2020).

4.2   Government Regulation

Complying with AI can help banks process an increasing array of regulations more efficiently, identify unusual behaviour within the staff, and detect suspected personal functioning. In marketing and sales, AI can prescribe specific items to clients and encourage with an even more accurate targeting of customers. In deposit and account management, AI can support commercial banks by lowering the cost of cash transactions and offering different services to its clients (Königstorfer & Thalmann, 2020).

4.3   Customer Satisfaction and Requirement

Techniques to detect fraud and theft will make transaction processing safer for customers. AI could be used to forecast cash usage and reduce operating costs for ATMs and branch networks (Königstorfer & Thalmann, 2020). Banks must supply collaborator platforms and services so that they could engage customers and take advantage of stakeholders' information and channel platforms to maximise their participation across various modes (such as the network, mobile app, industry, call centre, and smart devices). Example for this is ICICI Bank in India integrated WhatsApp (a common India social app) in financial services and within three months, pared back to one million users (Ojha, 2020) and Citibank has also employed machine learning and big data and invested more than $11 million in a new anti-money laundering system (Padhi, 2019).

5   CONCLUSION

Across the world, not just client sectors are embraced for machine learning and artificial intelligence, but they also turn the company model from procedure to platforms. The AI change increased the competitiveness of the company. The definition was historically extended to few roles in the organisation, but it was eventually confined to all of the other areas of operation. The Banking industry has entered AI in order to promote quicker customer service and to allow banks to better manage data and to build strategic plans. As per PwC report AI contributed 2 trillion dollar to GDP in the year 2018 and it will reach 15.7 trillion dollar by the year 2030 (Rao, 2020). As per Nasscom India GDP could reach to 450–500 billion dollar (The Hindu, 2020). In china the projected value of share of AI contributions to GDP has reached highest by 26.1% approximately followed by North America contributing to 14.5% (Statista, 2020). There is much more to do here in the banking industry as some queries cannot be processed through chatbots or controlled human robots. The involvement of AI needs to be encouraged by the Indian government to detect fraud in systems to narrow the space and explain market functions and improve customer satisfaction. The government must also create its own applications, as few companies such as Google, Alexa, Oracle, Siri and others are currently capturing the Indian market. India is able to promote its own applications for financial sectors and numerous other industries where AI is used as cloud-based data storage. Chugh and

Jaiswal (2020) the banking sector can regain customer confidence and satisfaction through AI, which can be easily assessed by technology and improve investment functions. If AI is implemented successfully in the banking sector, fraud and security problems will be minimised and much more visibility and realistic decisions will be made as it operates with the data entered.

References

Ataullah, A., & Le, H. (2006). Economic reforms and bank efficiency in developing countries: The case of the Indian banking industry. *Applied Financial Economics, 16*(9), 653–663. https://doi.org/10.1080/096031005004 07440.

Barthet, R. (2019). *European Banking Federation aisbl EBF position paper on AI in the banking industry.* www.ebf.eu.

Baruah, A. (2020, February 27). *AI applications in the Top 4 Indian Banks | Emerj.* Emerj. https://emerj.com/ai-sector-overviews/ai-applications-in-the-top-4-indian-banks/.

Bhushan, K. (2018, July 9). *Artificial Intelligence in Indian banking: Challenges and opportunities.* Livemint. https://www.livemint.com/AI/v0Nd6Xkv0nIN DG4wQ2JOvK/Artificial-Intelligence-in-Indian-banking-Challenges-and-op. html.

Biswas, S., Carson, B., Chung, V., Singh, S., & Thomas, R. (2020). *AI-bank of the future: Can banks meet the AI challenge?* McKinsey. https://www.mck insey.com/industries/financial-services/our-insights/ai-bank-of-the-future-can-banks-meet-the-ai-challenge.

Bughin, J., Seong, J., Manyika, J., Chui, M., & Joshi, R. (2018, September 4). *Notes from the AI frontier: Modeling the impact of AI on the world economy.* McKinsey. https://www.mckinsey.com/featured-insights/artificial-intellige nce/notes-from-the-ai-frontier-modeling-the-impact-of-ai-on-the-world-eco nomy.

Cam, A., Chui, M., & Hall, B. (2019, November 22). *Global AI Survey: AI proves its worth, but few scale impact.* McKinsey. https://www.mckinsey.com/featured-insights/artificial-intelligence/global-ai-survey-ai-proves-its-worth-but-few-scale-impact.

Chalimov, A. (2019, May 28). *Examining AI uses in banking & financial services.* Eastern Peak. https://easternpeak.com/blog/examining-ai-uses-in-banking-financial-services/.

Christensen, C. M., Hall, T., Dillon, K., & Duncan, D. S. (2016). Know your customers' "jobs to be done." *HBR.* https://hbr.org/2016/09/know-your-customers-jobs-to-be-done.

Chugh, N., & Jaiswal, S. (2020). *How AI is disrupting the banking sector in India*. SiliconIndia. https://technology.siliconindia.com/viewpoint/ceo-insights/how-ai-is-disrupting-the-banking-sector-in-india-nwid-9820.html.

Chui, M., & Malhotra, S. (2018, November 13). *AI adoption advances, but foundational barriers remain*. McKinsey. https://www.mckinsey.com/fea tured-insights/artificial-intelligence/ai-adoption-advances-but-foundational-barriers-remain.

Darshnik. (2019). *Cyber frauds in the Indian banking industry*. Legal Service India. http://www.legalserviceindia.com/legal/article-3073-cyber-frauds-in-the-indian-banking-industry.html.

Dietz, M., Khanna, S., Olanrewaju, T., & Rajgopal, K. (2016, February 1). *Cutting through the noise around financial technology*. Mckinsey. https://www.mckinsey.com/industries/financial-services/our-insights/cutting-thr ough-the-noise-around-financial-technology.

Digalaki, E. (2019, December 18). *The impact of artificial intelligence in the banking sector & how AI is being used in 2020*. Business Insider. https://www.businessinsider.in/finance/news/the-impact-of-artificial-intelligence-in-the-banking-sector-how-ai-is-being-used-in-2020/articleshow/728608 99.cms.

DR, B. (2020a, July 9). Banking on AI: The time is ripe for Indian banks to embrace artificial intelligence. *The Financial Express*. https://www.financ ialexpress.com/opinion/banking-on-ai-the-time-is-ripe-for-indian-banks-to-embrace-artificial-intelligence/2017963/.

DR, B. (2020b, July 9). *Banking on AI: The time is ripe for Indian banks to embrace artificial intelligence*. Financial Express. https://www.financialexp ress.com/opinion/banking-on-ai-the-time-is-ripe-for-indian-banks-to-emb race-artificial-intelligence/2017963/.

Ferreira, P., Teixeira, J. G., & Teixeira, L. F. (2020). Exploring service science. In H. Nóvoa, M. Dra˘goicea, & N. Kühl (Eds.), *10th International Conference, IESS 2020: Vol. 377 LNBIP* (pp. 202–213). Springer Nature. https://doi.org/10.1007/978-3-030-38724-2_7.

Frankenfield, J. (2020, March 13). *Artificial Intelligence (AI) definition*. Investo-pedia. https://www.investopedia.com/terms/a/artificial-intelligence-ai.asp.

Garvey, J., & Burns, P. (2020). *Financial services technology 2020 and beyond: embracing disruption*. http://www.pwc.com/gx/en/financial-services/.

Ghosh, M. (2017, September 5). *SBI Launches SBI Intelligent Assistant (SIA)— Chatbot for customer care*. Trak. https://trak.in/tags/business/2017/09/05/sbi-chatbot-sia/.

Guardians, N. (2019). *All about digital banking fraud prevention— NetGuardians*. Net Guardians. https://netguardians.ch/digital-banking-fraud/.

Gujral, R. (2020, April 29). *How banking benefits from AI and voice technology—AI Business*. AI Business. https://aibusiness.com/document.asp?doc_id=761287.

Gupta, J. (2020, December 22). *The influence of artificial intelligence on the banking industry*. Wire19. https://wire19.com/influence-of-artificial-intelligence-on-banking-industry/.

Harman, R. (2019, July 11). *The growing impact of AI on the banking industry | Onlim*. Onlim. https://onlim.com/en/the-growing-impact-of-ai-on-the-banking-industry/.

Hwang, H. G., Ku, C. Y., Yen, D. C., & Cheng, C. C. (2004). Critical factors influencing the adoption of data warehouse technology: A study of the banking industry in Taiwan. *Decision Support Systems, 37*(1), 1–21. https://doi.org/10.1016/S0167-9236(02)00191-4.

Ibbitson, C. (2020, December 10). *Using AI to transform the customer experience in banking*. Finextra. https://www.finextra.com/blogposting/19634/using-ai-to-transform-the-customer-experience-in-banking.

Jadhav, S., He, H., & Jenkins, K. (2016). An academic review: Application of data mining techniques in finance industry. *International Journal of Soft Computing and Artificial Intelligence, 4*(1), 79–95. http://iraj.in/journal/IJSCAI/volume.php?volume_id=258.

Jewandah, S. (2019). How artificial intelligence is changing the banking sector—A case study of top four Commercial Indian banks. *International Journal of Management, Technology And Engineering, 8*(7), 525–530. http://ijamtes.org/gallery/66.julyijmte-711.pdf.

Kaur, N., Sahdev, S. L., Sharma, M., & Siddiqui, L. (2020). Banking 4.0: "The influence of artificial intelligence on the banking industry & how AI is changing the face of modern day banks." *International Journal of Management, 11*(6). https://doi.org/10.34218/ijm.11.6.2020.049.

Kaya, O. (2019, June 4). *Artificial intelligence in banking: A lever for profitability with limited implementation to date*. DBresearch. http://www.dbresearch.com/PROD/RPS_EN-PROD/HIDDEN_GLOBAL_SEARCH.alias.

Kharod, S. (2020, August). *The 5 major benefits of AI in the banking and finance sector | chatbots journal*. Chatbots Journal. https://chatbotsjournal.com/the-5-major-benefits-of-ai-in-the-baking-and-finance-sector-c1451562f41a.

Khurshid, A. (2020). *Banking on Artificial Intelligence (AI)—Wipro*. Wipro. https://www.wipro.com/business-process/why-banks-need-artificial-intelligence/.

Kumar, G., Muckley, C. B., Pham, L., & Ryan, D. (2019). Can alert models for fraud protect the elderly clients of a financial institution? *European Journal of Finance, 25*(17), 1683–1707. https://doi.org/10.1080/1351847X.2018.1552603.

Kumar, A. (2017, September 14). *How artificial intelligence is impacting the service industry.* Yourstory. https://yourstory.com/2017/09/artificial-intell igence-impacting-itsm?utm_pageloadtype=scroll.

Kvamme, H., Sellereite, N., Aas, K., & Sjursen, S. (2018). Predicting mortgage default using convolutional neural networks. *Expert Systems with Applications, 102,* 207–217. https://doi.org/10.1016/j.eswa.2018.02.029.

Königstorfer, F., & Thalmann, S. (2020). Applications of Artificial Intelligence in commercial banks—A research agenda for behavioral finance. *Journal of Behavioral and Experimental Finance, 27,* 100352. https://doi.org/10. 1016/j.jbef.2020.100352.

Latimore, D. (2018). *Artificial Intelligence in banking where to start?* https:// cdn2.hubspot.net/hubfs/2392451/MarketingAssets/ArtificialIntelligencein Banking-WheretoStart.pdf.

Maheshwari, A. (2020, December 22). Banking of tomorrow with human and artificial intelligence. *Daily Host News.* https://www.dailyhostnews.com/ban king-of-tomorrow-with-human-and-artificial-intelligence.

Marous, J. (2020). *Meet 11 of the most interesting chatbots in banking.* The Financial Brand. https://thefinancialbrand.com/71251/chatbots-banking-tre nds-ai-cx/.

Martens, D., Provost, F., Clark, J., & Junqué de Fortuny, E. (2016). MIS quar-terly. *MIS Quarterly, 40*(4), 869–888. https://misq.org/mining-massive-fine-grained-behavior-data-to-improve-predictive-analytics.html.

Maru, P. (2017, October 17). ICICI Bank's AI chatbot iPal empowers customers with information and financial services, IT News, ET CIO. *India Times.* https://cio.economictimes.indiatimes.com/news/enterprise-services-and-app lications/icici-banks-ai-chatbot-ipal-empowers-customers-with-information-and-financial-services/61118452.

McGlasson, L. (2010, May 17). *5 Tips to reduce banking fraud—BankInfoSecu-rity.* Bank Info Security. https://www.bankinfosecurity.com/5-tips-to-reduce-banking-fraud-a-2534.

Mejia, N. (2020, March 10). *AI-based fraud detection in banking—Current applications and trends.* Emerj. https://emerj.com/ai-sector-overviews/artifi cial-intelligence-fraud-banking/.

Michałowska, M. (2020, April 16). *How Artificial Intelligence is influencing the banking sector.* Learn.G2. https://learn.g2.com/ai-in-banking.

Nandi, S. (2020, December 8). *Fraud and financial crime reduction in banking with AI—RTInsights.* RTI Sights. https://www.rtinsights.com/fraud-and-fin ancial-crime-reduction-in-banking-with-ai/.

Ngai, E. W. T., Hu, Y., Wong, Y. H., Chen, Y., & Sun, X. (2010). The appli-cation of data mining techniques in financial fraud detection: A classification framework and an academic review of literature. *Decision Support Systems, 50,* 559–569. https://doi.org/10.1016/j.dss.2010.08.006.

Ojha, S. (2020, July 7). *ICICI Bank crosses 1 million users on WhatsApp platform.* Livemint. https://www.livemint.com/money/personal-finance/icici-bank-crosses-1-million-users-on-whatsapp-platform-11594098935957.html.

Padhi, U. (2019, August 21). *AI in banks: Risks and opportunities.* Fintechnews. https://www.fintechnews.org/ai-in-banks-risks-and-opportunities/.

Palepu, A. R. (2020, August 25). *RBI Annual Report 2019–20: Bank frauds more than double.* Bloomberg Quint. https://www.bloombergquint.com/bus iness/rbi-annual-report-2019-20-bank-frauds-more-than-double.

Parayil, M. (2018, August 9). *Chatbot in banking—Examples, best use cases and the future.* Haptik. https://www.haptik.ai/blog/chatbots-in-banking-exa mples-best-usecases-future/.

PTI. (2019, December 24). Total frauds at banks rise 74 per cent to Rs 71,543 crore in 2018–19: RBI. *The Economic Times.* https://economictimes.indiat imes.com/industry/banking/finance/banking/total-frauds-at-banks-rise-74-per-cent-to-rs-71543-crore-in-2018-19-rbi/articleshow/72957892.cms?fro m=mdr.

Rao, A. (2020). *PwC's global Artificial Intelligence study: Sizing the prize.* PwC. https://www.pwc.com/gx/en/issues/data-and-analytics/public ations/artificial-intelligence-study.html.

Rivera, K., Rohn, C., Donker, J., & Butter, C. (2020). *PwC's global economic crime and fraud survey 2020.* PwC. https://www.pwc.com/gx/en/services/ forensics/economic-crime-survey.html.

Russell, S., & Norvig, P. (2013). Artificial Intelligence: A modern approach, 3rd Edition | Pearson. In *Pearson/Prentice Hall* (3rd ed.). Publisher. https:// www.pearson.com/us/higher-education/program/Russell-Artificial-Intellige nce-A-Modern-Approach-3rd-Edition/PGM156683.html.

Sharma, S. (2018, May 22). *10 AI-powered virtual assistants making banking easier for everyday consumers.* Yourstory. https://yourstory.com/2018/05/ consumer-banking-ai-virtual-assistants.

Shu, W., & Strassmann, P. A. (2005). Does information technology provide banks with profit? *Information and Management, 42*(5), 781–787. https:// doi.org/10.1016/j.im.2003.06.007.

Statista. (2020, August). *Worldwide: Projected AI contribution to GDP by region 2030.* Statista. https://www.statista.com/statistics/1042325/worldwide-pro jected-ai-contribution-gdp-by-region/.

Stephanie. (2019, May 16). *Voice assistants, chatbots, and AI—Don't miss out!.* Onlim. https://onlim.com/en/voice-assistants-chatbots-and-ai-dont-miss-out/.

The Hindu. (2020, August 18). Data, AI may add up to $500 bn to GDP by 2025. *The Hindu.* https://www.thehindu.com/business/data-ai-may-add-up-to-500-bn-to-gdp-by-2025/article32388142.ece.

Tobback, E., & Martens, D. (2017). Retail credit scoring using fine-grained payment data. In *Journal of the Royal Statistical Society* (No. 2017011; Series A (Statistics in Society)). University of Antwerp, Faculty of Business and Economics. https://ideas.repec.org/p/ant/wpaper/2017011.html.

Trehan, R. (2020). *How AI is transforming Fraud Prevention in Banking and Finance*. Deltec Bank & Trust. https://www.deltecbank.com/2020/05/11/how-ai-is-transforming-risk-in-finance-and-banking/?locale=en.

Turing, A. M. (1950). Computing machinery and intelligence. In *Mind* (Vol. 49, pp. 433–460). Springer. https://doi.org/10.5603/ARM.a2017.0044.

Uzialko, A. (2019, April 22). *How Artificial Intelligence is transforming business—businessnewsdaily.com*. Businessnewsdaily. https://www.businessn ewsdaily.com/9402-artificial-intelligence-business-trends.html.

Walch, K. (2020, April 5). *Why AI Is transforming the banking industry*. Forbes. https://www.forbes.com/sites/cognitiveworld/2020/04/05/why-ai-is-transforming-the-banking-industry/#53e81fb67dd6.

Zhang, S., & Woo, R. (2018, February 1). *Alibaba-backed online lender MYbank owes cost-savings to home-made tech | Reuters*. Reuters. https://www.reuters.com/article/us-chinabanking-mybank/alibaba-backed-online-lendermybank-owes-cost-savings-to-home-made-techidUSKBN1FL3S6.

Zhou, C. (2020, May 12). *How can AI fraud detection help the banking industry?» DataVisor*. Datavisor. https://www.datavisor.com/blog/how-can-ai-fraud-detection-help-the-banking-industry/.

Understanding the Emerging Role and Importance of Robo-advisory: A Case Study Approach

Shreshtha Singhvi

1 Introduction

The investing panorama is dynamic and in this continuously evolving environment there are a plethora of options available to an investor today to choose from and invest in. The investor can buy and trade in stock market, commodity market, exchange-traded funds (ETFs) and invest with mutual funds to list a few. The investor can buy and trade in stock market, commodity market, exchange-traded funds (ETFs) and invest with mutual funds to list a few.

Trading in stock market allows the investor to partake in the company's success via upward change in the stock prices and receipt of dividends. Equity shareholders also enjoy voting rights at annual meetings of the company and gain a claim on assets of the company in the event of company's liquidation. However, Preference shareholders do not possess voting rights but receive preference over equity shareholders at the time of payment of dividend by the company.

Trading in commodity markets in some countries pre-dates trading in stock and bond market. In ancient history, ascent of many dynasties could be directly attributed to their ability to device effective trading mechanism for commodities. Commodity markets, presently are regulated

S. Singhvi (✉)
Commercial Banking Group IndusInd Bank, Mumbai, India

© The Author(s), under exclusive license to Springer Nature Singapore Pte Ltd. 2021
N. R. Al Mawali et al. (eds.), *Fourth Industrial Revolution and Business Dynamics*, https://doi.org/10.1007/978-981-16-3250-1_3

by various exchanges across the globe. Corn, wheat, sugar, rice, pulses, coffee, metals, oil, natural gas, etc. are some of the commodities traded in the commodity markets globally.

Exchange-Traded Funds (ETFs) have gained traction since their conception in the mid-1990s. They are similar to mutual funds and can be traded 24×7 on stock exchanges. They function in the same way as stocks on a stock exchange and their value fluctuates throughout the trading day. ETFs generally track an underlying stock index such as S&P500 or any other index, with which, an ETF wants to align itself. ETFs are inclusive of wide range of classes such as emerging markets, business sectors, commodities to cite a few.

Mutual funds enable individual investors to pool their money collectively to purchase securities through funds which are managed and catered for by a dedicated fund manager. The role of a fund manager is to allocate and distribute the money invested by individual investors into stocks, bonds and other securities. Investors in return are charged a management fee by the fund manager for services provided.

Traditionally, the investors have either relied on their own studies, recommendations from their agents/brokers/friends and print or visual media for their investments in the markets. With the advent of the fourth industrial revolution, new methods of analysing the markets and advising the investors have emerged. One such method is called robo-advisory investing. Robo-advisory is a function of machine learning technique and an application of artificial intelligence in the field of finance. It is an online investment management platform rendering automated financial advice to investors as per their requirements. The technology platform learns and ameliorates from the interactions without being categorically programmed. The machine learning technique helps the platform to access data and use the data collected to learn for itself. The investor's input is translated into investment logic such as liquidity adequacy, risk appetite, purpose adequacy to name a few. Majority of the portfolios are ETF based as per the investor's requirements.

Robo-advisory platforms render advice based on the inputs provided by the investor, seeking financial advice. They generate a custom-made portfolio and investment recommendations tailored to the requirements of the investor. The complete premise of the aforesaid technology is based on the fact that, this unique technology platform can render unbiased, rational and objective financial advice at an economical cost when compared to its traditional counterparts.

This chapter aims to comprehend various robo-advisory models at present. Further, it shall also discuss the road ahead and future market scenario for robo-advisors. The chapter also encompasses the features and functions of robo-advisors by means of a case study presented in the chapter. This study is based on personal experience and secondary data available in the public domain. The secondary data comprises of previous studies, various reports, white and working papers.

2 Literature Review

Dellaert (2017) suggest that a robo-advisor refers to a robotic assistance that positions, or matches clients to monetary items on a customised premise and sells customised products to its clientele. Further, the platform also aids in educating its clients via sharing knowledge on investing. Their study depicts few common features of robo-advisors namely delivering of comprehensive digitised services, conducting automated portfolio rebalancing, implementing indexation or passive management strategies and personalised investor goals (Dellaert, 2017).

Kane (2014) has made a relative investigation between man and machine. She has contrasted robo-advisors with human guide on different parameters like price, portfolio size and other variants. Robo-advisors levy much lower fees as compared to other investment alternative (Kane, 2014). With respect to portfolio size, robo-advisors are more efficient as compared to traditional financial advisors as they are equipped with faster speed to compute the data and flexibility to incorporate any change required in the asset allocation. The output rendered by a robo-advisory is unbiased and indifferent to any opinion or conflict of interest as the process of data feeding is absolutely transparent. However, there is an area where human advisors supersede the robo-advisory platforms, wherein, robo-advisory platforms find difficulty in comprehending the personality traits and priority of goals of the investors in a dynamic and ever-changing environment (Moewes et al., 2011).

Huxley and Kim (2016) have analysed 17 robo-advisors and stated that the recommendations made by robo-advisory platforms for equity portfolios are more suitable in the short run as compared to long run (Huxley & Kim, 2016). Robo-advisors typically attempt to reduce short-term volatility. Further, the report emphasises on clients planning for their retirement, and have investment horizons of more than 10 years, they identify a conflict between true investment goals and recommendations.

This conflict is linked back to the quality of the questionnaires that enable the robo-advisors to ascertain the clients' risk preferences. Huxley and Kim (2016) augment a critical issue—Whether or not the questionnaires of robo-advisory platforms are valid and precise from a psychometric view of an investor. The report highlighted that the questionnaires did not adhere to psychometric standards as they tend to be too brief. Considering the gaps in the report, in the opinion of the author, the future of financial advisory services will evolve into a hybrid model, where analysis is done by programmed algorithms but the procuring, retaining, motivating and reassuring of clients will be done by human counterpart (Roszkowski, 2005).

3   Working of Robo-advisors

The robo-advisory platform commences the advisory process by understanding individual investor's goal, risk appetite and time horizon. Subsequently an appropriate goal-based investment strategy is offered by the robo-advisor curated to the needs and requirement of the said investor. The goals of investment for an investor can include funds for retirement, savings to cover larger expenses, setting up of an emergency fund, funds for children's future education to name a few.

To gauge the goals of its clients (investors), a set of objective and subjective questions in the form of a questionnaire is presented to the clients. The aforesaid questionnaire aids in evaluating the readiness and risk tolerance levels of the clients. The objective questions help in understanding clients' income levels, retirement period and other defined criteria. The subjective questions aid in understanding clients' behaviour towards movements, fluctuations and variances in the market (Lam, 2016). For a simple and cost-efficient process, the assessment of the clients is conducted by using short questionnaires via robo-advisory platforms.

With the aid of automated algorithms, robo-advisory platforms suggest how to allocate funds across varied types of asset classes. These algorithms are largely driven by modern portfolio theory (Vukovic & Bjernes, 2017). The asset allocation model, used industry wide stems its roots from portfolio selection framework of Markowitz. The model states that in order to obtain an optimal portfolio, either the expected return must be maximum at a certain level of risk or the risk should be minimised at a given level of expected return (Markowitz, 1952). Generally, the portfolios curated by robo-advisors augment the risk component by enhancing the equity to debt ratio within each asset class, thereby, investing in riskier asset classes (eg: increasing the share of investment in municipal bonds in comparison to government bonds or increasing the asset base in emerging markets rather than developed markets).

In robo-advisory managed portfolios, optimization occurs after considering clients' investment goals and the desired risk levels. This can be understood with the help of a scenario where for a given level of risk, the objective is to generate income to meet expenditure or accumulate long-term saving. The algorithms can be structured to continuously monitor portfolios and detect deviations and variances from the targeted risk coupled with recommendation of initial allocation of funds. The portfolio is automatically rebalanced when deviations occur and are spotted. For instance, when funds are invested in both equity and bond asset classes and if over time value of equity increases faster than the value of bond thereby increasing the share of equity in the portfolio. As the share of equity increases the risk augments and the portfolio will rebalance itself by selling equity in order to align with the goal and risk appetite of the clients. Rebalancing can also take place when the clients' change their risk tolerance levels and /or investment goals. In order to gain diversification at lower costs, robo-advisors majorly invest in index funds and exchange-traded funds (ETFs).

The investment instruments deployed by robo-advisory platforms tend to abide to an array of securities or an index. Generally, the observations of outcomes witness that, investors tend to achieve a "market portfolio" while incurring minimal trading costs only by acquiring a handful of funds. Additionally, by passively holding funds, investors need not engage in active monitoring and trading and thus can achieve reduced transactions costs.

The industry trend depicts that the robo-advisory platforms work on a rather conservative approach. Robo-advisors can combine the computing and judgement resources of machine with the human brain (Phoon & Koh, 2018). They offer funds with wide coverage, long established operating history, market liquidity and good performance over time (Kaya, 2017). The working of robo-advisors is depicted in Fig. 1.

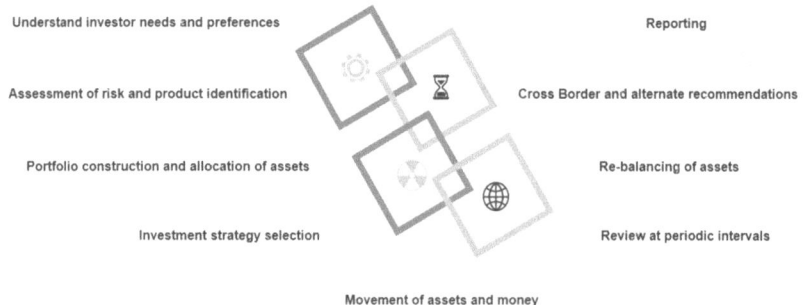

Fig. 1   Working of robo-advisors (*Source* Prepared by author)

4   Robo-advisory Models

Innovative technological developments are sprouting in the field of finance thereby making a revolutionary shift in the current marketplace for investment products and services rendered to investors. Currently, advice for wealth management is provided by traditional methods as well as robo-advisory and hybrid models. These models are summarised in Fig. 2.

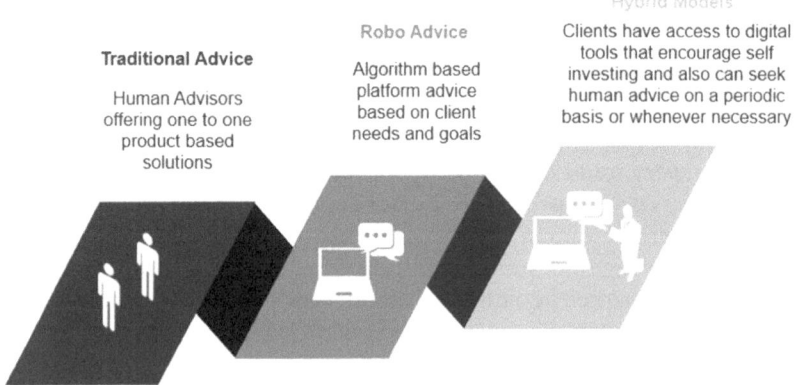

Fig. 2   Robo-advisory models (*Source* Prepared by author)

There are majorly two kinds of robo-advisory models shaping up in the industry and some identifiable features of these models are discussed below:

a. Fully—automated model

(i) This model possesses the ability to reach a larger target group of investor segment. It is mainly used to reach masses.
(ii) This model lacks live advisors or human element.
(iii) There is little or no scope for personalised financial advice to start with. After on- boarding, with the historical data personalisation can be possible.
(iv) The investible asset classes are limited.
(v) Lower fees charged to investors result in improved profit margins.

b. Hybrid model

(i) This model operates with a limited group of investors. This model leverages robots for advice and human element for management of clients.
(ii) This model provides the facility of a live advisor to clients as per the needs and requirements of clients.
(iii) From on-boarding to the exiting stage of a client, customised experience can be arranged under this model.
(iv) The advice rendered under this model is based on varied products and currencies available across markets.
(v) Investors who are investing for a long period of time are open to paying a fee for personalised services.

5   Market Landscape of Robo-advisory

Today, Assets Under Management (AUM) with the top 10 robo-advisors in the world approximate to US$260,000 m. The largest robo-advisors in terms of assets under management are Vanguard (US$140,000 m), followed by Schwab Intelligent Portfolios (US$41,000 m) and Wealth-front (US$20,000 m) (Table 1).

Robo-advisors are spreading their reach to all parts of the world. The United States of America has the maximum number of robo-advisors, while Europe houses over 70 robo-advisors, with 5 of them managing

Table 1   Top 10 robo-advisors

Sl. No	Robo-advisors	AUM (in million US$)
1	Vanguard Personal Advisory Service	140,000
2	Schwab Intelligent Portfolios	41,000
3	Wealthfront	20,000
4	TD Ameritrade Essential & Selective Portfolios	19,900
5	Betterment	18,000
6	Personal Capital	10,000
7	*Etrade Core Portfolios*	6200
8	WealthSimple	3000
9	FutureAdvisor	1200
10	M1 Finance	500

Source Data from Carey (2019), table prepared by author

more than €100 m. Emerging economies are starting to gain traction too with their own robo-advisors coming up. For instance, Asia is growing fast in the robo-advisory space. Presence of robo-advisory platforms is seen in China (Mainland), Hong Kong SAR, China, Japan, India, Thailand, Singapore, and Vietnam among other economies. This spread is driven by the growing middle class in these countries and evolving technological connectivity. Further, Africa and Latin America together are a home to six robo-advisors (Burnmark, 2017). In the coming years robo-advisors' presence is expected to spread in size and scope.

The approximate value of AUM with the robo-advisors in 2019 is about US$980,541 m and is anticipated to grow at a CAGR of 27.0% and reach to approximate US$2,552,265 m by 2023 (Fig. 3). The number of users using the robo-advisors was approximately 13,105 (in '000s) in 2017 and is projected to grow up to 147,018 (in '000s) in 2023 (Fig. 4).

6   Robo-advisory Landscape in Middle East

The dynamics of wealth management in the middle east is going through a radical alteration as the regulators in the region are embracing robo-advisory platforms. The robo-advisory market in middle east is predicted to grow at a compound annual growth rate (CAGR) of ~55% and the approximate value of AUM will be US$3800 m by 2023. The governments of these countries are progressively adopting policy measures to encourage the use of robo-advisors for wealth management. Presently,

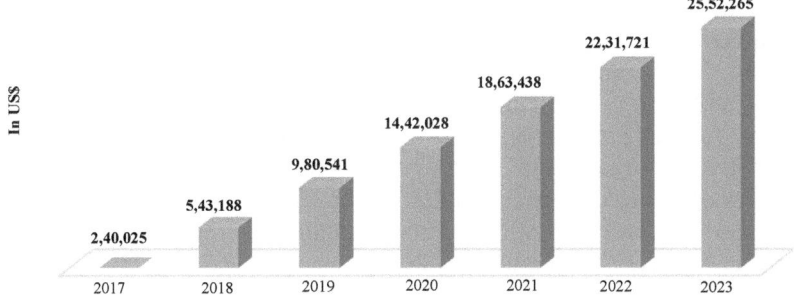

Fig. 3 Projections of assets under management with robo-advisors (*Source* Data from Statista September (2019), figure prepared by author)

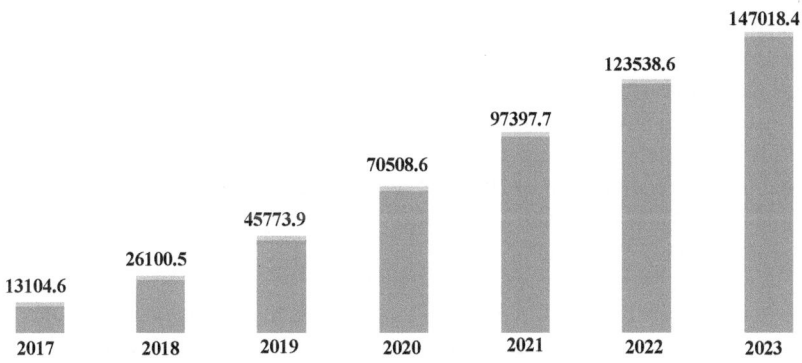

Fig. 4 Users of robo-advisor platforms (*Source* Data from Statista September, 2019 (figures in '000s), figure prepared by author)

majority of the robo-advisory platforms in middle east are housed in United Arab Emirates and Saudi Arabia. Dominant players in this market space are Sarwa, Wahed Invest and Haseed Investing. Hybrid robo-advisory and pure robo-advisory models are more dominant in the middle eastern markets.

The governments of these countries are progressively adopting policy measures to encourage the use of robo-advisors for wealth management. To enhance its position as a digital finance hub and take full advantage of advances in IT, the authorities of Bahrain are adopting measures to foster financial inclusion. The Central Bank of Bahrain has issued a set of directives regulating the role of robo-advisors in the country (cpifinancial, 2019). Similarly, a regulatory framework for robo-advisors has been published by the Financial Services Regulatory Authority (FSRA) of the Abu Dhabi Global Market (ADGM) (sarwa.co, 2019). Capital Market Authority of Saudi Arabia has approved testing of robo-advisory services of Waheed Capital and Haseed Investing. Both these platforms will also be allowed to offer automated online discretionary investment management services (Goncalves, 2019).

7   Case Study on Betterment

Post financial crisis of 2008, a large increase in the investor class was witnessed who were tech savvy professionals and preferred digital advice for management of investments there by leading to the evolution of robo-advisors. Robo-advisors bridged the existing lacuna of those times when the well-established wealth management firms required a large minimum sum to begin investing. Unlike the large firms, robo-advisory platforms required a small minimum amount of money to start investing. The well-established financial advisory firms also charged a high amount of fees for their services rendered while robo-advisory charged a much lesser fees compared to the former.

An MBA graduate from Columbia Business School, Jon Stein and a lawyer from NYU School of Law, Eli Broverman together established a company called Betterment in the year 2008 after the crisis at New York. This company is a first generation robo-advisor with approximately US$18,000 m assets under management. The building of first online platform for Betterment was started by Stein alongside his roommate Sean Owen, who was a software engineer with Google in 2008. They used a Java application and MySQL database on Apache Tomcat servers with an Adobe Flash and Flex-based front-end design. Stein's friend, Polina Khentov supplied them with the Initial prototype designs. Eli Broverman, then a securities lawyer was brought on board to comply with the financial regulations associated with starting a financial company. Betterment's next addition to the team was Ryan O'Sullivan, a seasoned entrepreneur, to develop Betterment's broker-dealer operation (Stein, 2016).

From the period 2008–2010, the founding members of the company worked continuously to prepare Betterment for launch. In the aforesaid period, the company also gained Financial Industry Regulatory Authority

(FINRA) membership. Anthony Schrauth, a former colleague of Stein's joined the company as the chief of product officer in 2009, and Owen was replaced by Kiran Keshav of Columbia University's Center for Computational Biology. O'Sullivan left his role as president in 2010. On 7 April, 2009, Betterment LLC transformed itself to Delware Corporation. The company bagged the award of "Biggest New York Disruptor" after its initial launch at TechCrunch Disrupt New York in June 2010 (Stein, 2016, Betterment.com). After the launch, 400 new clients had signed up to invest their money with Betterment.

Betterment began its journey with the initial product portfolio of taxable investment and retirement accounts. Since then, the company has steadily increased its portfolio by addition of new products like smart beta, socially responsible investing, and fixed income centred investments. At the time of setting up of an account, the client is asked to choose his goals from various categories namely savings, education, safety net, general investing, retirement or major purchase. When a client on-boards with Betterment, Betterment begins the on-boarding process by presenting a questionnaire to its client. The questions in the questionnaire are created to gauge the client's investment goals, risk appetite and timeline.

After receiving the responses to the questionnaire, the advisor then builds a custom-made portfolio for the client to match his/her requirements. As a thumb rule, the advisor allocates the money of the client in three distinct categories:

a. Retirement: This category focusses on retirement funds, tax sheltered plans and long-term savings.
b. Build Wealth: This category focusses on medium term goals such as client's children's education, a particular goal that the client needs funds for, a house or investing in general.

For instance, if a client is 35 years old with an annual income of US$100,000. After perusing the responses on the questionnaire, below would be the recommendations from the advisor (using the thumb rule) under each category.

a. Safety Net: The recommendation is conservative as it suggests to invest 40% in shares and 60% in bond.

b. General Investing (Traditional IRA): The recommendation is aggressive as it suggests to invest 90% in shares and 10% in bonds.
c. Build Wealth: The recommendation is aggressive as well as it suggests to invest 90% in shares and 10% in bonds.
d. House Fund: The recommendation is aggressive as it suggests to invest 60% in shares and 40% in bond.

Some of the features of Betterment include:

a. Minimum Initial Investment—Betterment is one of the few robo-advisors which open client's account without any minimum amount. To begin investing the client can start with $100 per month.
b. Product Portfolio—Betterment offers individual and joint taxable accounts; traditional, Roth, rollover, and SEP IRAs; trusts and non-profit accounts.
c. Protection—All the accounts of clients created on Betterment's interface are Securities Investors Protection Corporation (SIPC) protected up to US$500,000 in securities and US$250,000 cash.
d. Security—It has a two-step security process. The first step requires the client to input the access code and the second step requires the client to input a verification code sent via text, email or call as per the client's comfort.
e. Fees—Betterment offers two categories of plans to its clients:

(i) Digital—Here annual fee is 0.25% on the initial US$2 million and 0.15% on excess balances.
(ii) Premium—Here annual fee is 0.40% on the initial US$2 million and 0.15% on excess balances.

f. Betterment Retirement Guide—The advisor has a special tool setup in the interface which is dedicated for the clients to help them plan their retirement financially. For instance, the tool calculates the amount required by them at the time of retirement basis their current age and present level of savings.

Betterment diverts the clients' portfolio into 14 asset classes. Each asset class is represented by a single ETF. The asset classes are divided into six stocks and eight bond asset classes. The stock asset class includes US

Total stock market, US Value stocks in large, mid and small cap as well as international developed and emerging market stocks. The bond asset class includes US high quality, municipality, inflation-protected, high-yield corporate, short-term treasury and short-term investment grade bonds. Besides, these international developed and emerging market bonds are also included in this bond asset class.

8 Advantages of Robo-advisors

In the era of digitised services, robo-advisor is an automated wealth management tool that uses algorithms to create investment strategies. Some of the advantages of robo-advisors are given below:

a. User friendly—To begin investing with a robo-advisor is simple and fast. A questionnaire containing a set of questions, designed to assess the risk appetite, goals and financial position, is presented at the time of account setup; based on the inputs the software then generates a portfolio which the user can approve or adjust according to his or her liking.

b. Economical service—Before the evolution of robo-advisors, traditional wealth management services charged a minimum of 1% of AUM from the investors for their investment management services. Nowadays, the robo-advisory platforms charge 0–0.25% of AUM thereby reducing the cost of services drastically.

c. Minimal investment—Traditionally, wealth managers required a sizeable sum to begin investing with them where as robo-advisors require as low as $500 or $1000 to begin investing. Robo-advisors like betterment require no minimum balance to open an account with it.

d. Rebalancing—Robo-advisors continuously monitor the portfolio and balance the asset allocation in the portfolio between the asset classes selected to align the portfolio as per the user's requirements and preferences. Rebalancing helps to keep the percentage invested in asset classes such as stocks, bonds and other funds constant and in line with the investor's expectations.

Robo-advisors have remarkably changed the investment setup globally. These advisors have not only broadened the investment horizon but have attracted investors from all sections of the society.

9   SCOPE AND FUTURE OF ROBO-ADVISORS

Robo-advisory is a fast-growing application of Fintech to provide solutions for emerging issues of asset and wealth management. Using various kinds of models, these robo-advisors are providing solutions which are not only pocket friendly but are also unbiased in nature. Acting as efficient asset allocators these advisors are keeping the best interests of client in sight. Due to client convenience and innovative approach, these modern-day services are becoming popular amongst all sections and age groups. As such asset and wealth management in the future will be a combination of services rendered by robotics, automation and traditional (in person) advisory (Sairam, 2018). Along with their existing role, robo-advisors will increasingly contribute to financial inclusion in future with new applications in the field of information technology, data analysis and custom-made solutions to arrive at a holistic investment strategy. This will not only enrich the financial landscape but will also make it more inclusive to equitably benefit those sections of society that are unbanked or underbanked. As such the robo-advisors in future will have a prominent place in the financial ecosystem.

REFERENCES

Burnmark. (2017). *Digital wealth-robo investing*. Burnmark.
Carey, T. W. (2019). *Robo-advisors 2019: Still waiting for the revolution*. Invest opedia.com.
cpifinancial. (2019, April 1). *Robo-advisors set to shake up region's asset management industry*. Retrieved from cpifinancial https://www.cpifinancial.net/bme/news/robo-advisors-set-to-shake-up-regions-asset-management-ind ustry.
Dellaert, B. G. (2017). Regulating robo advice across the financial services industry. *Iowa Law Review, 713–750*.
Fisch, L. (2018). The emergence of the robo-advisor. In *Wharton Pension Research Council Working*.
Goncalves, P. (2019, July 31). *Saudi Arabia allows two firms to test robo-advisory services*. Retrieved from International Investment https://www.internati

onalinvestment.net/news/4003369/saudi-arabia-allows-firms-test-robo-adv
isory-services.

Huxley, S. J., & Kim, J. Y. (2016, September 12). The short-term vature of robo
portfolios. *Advisor Perspectives,* 1–8.

Kane, L. (2014, July 22). Robo-advisors vs. financial advisors: Which is better
for your money? *Business Insider.*

Kaya, O. (2017). *Robo-advice—A true innovation in asset management.* Bank
Deutsche.

Lam, J. W. (2016, April 4). *Robo-advisors: A portfolio management perspective.*
Retrieved from Yale University http://economics.yale.edu/sites/default/
files/files/Undergraduate/Nominated%20Senior%20Essays/2015-16/Jon
athan_Lam_Senior%20Essay%20Revised.pdf.

Markowitz, H. (1952). Portfolio selection. *The Journal of Finance,* 77–91.

Moewes, T., Puschmann, T., & Rainer, A. (2011). Service-based integra-
tion of IT innovations in customer-bank-interaction. In *Wirtschaftinformatik
Proceedings* (p. 102). Zurich: Proceedings der 10. Internationalen Tagung
Wirtschaftsinformatik (WI2011).

Phoon, K. P., & Koh, C. C. (2018). Robo-advisors and wealth management. *The
Journal of Alternative Investments,* 79–94.

Roszkowski, D. G. (2005). Insights from psychology and psychometrics on
measuring risk tolerance. *Journal of Financial Planning,* 66–77.

Sairam, M. V. (2018). Robo advisory in wealth management. *TCS BaNCS
Research Journal,* 42–45.

sarwa.co. (2019, August 2019). *What is robo advisor.* Retrieved from sarwa.co
https://www.sarwa.co/blog/what-is-a-robo-advisor.

Stein, J. (2016, July 20). *The history of betterment: Changing an industry.*
Retrieved from Betterment.com https://www.betterment.com/resources/
the-history-of-betterment/.

Vukovic, A., & Bjernes, L. (2017). *Automated advice: A portfolio management perspective on robo-advisors.* Trondheim: Norwegian University of
Science and Technolgy. https://ntnuopen.ntnu.no/ntnu-xmlui/bitstream/
handle/11250/2473732/17822_FULLTEXT.pdf?sequence=1&isAllowed=y.

Sectoral Impact

Healthcare Governance in the 4th Industrial Revolution: Leading Through Patient-Centeredness and Empowerment in the UAE

Virginia Bodolica and Martin Spraggon

1   INTRODUCTION

In the context of the 4th industrial revolution, giving patients a greater voice and more power in their interaction with suppliers of healthcare became a national priority in many countries around the globe (Dent & Pahor, 2015; Moretta Tartaglione et al., 2018; Snyder & Engstrom, 2016; Tofan et al., 2013a). Patient empowerment (PE) emerged as a focal element in governmental efforts to achieve patient-centeredness for inducing enhanced self-care efficacy, increased medication adherence, optimized resource usage, and improved citizens' wellbeing (Bodolica

V. Bodolica (✉)
The Said T. Khoury Chair of Leadership Studies, School of Business Administration, American University of Sharjah, Sharjah, UAE
e-mail: vbodolica@aus.edu

M. Spraggon
School of Business and Quality Management, Hamdan Bin Mohammed Smart University, Dubai, UAE
e-mail: m.spraggon@hbmsu.ac.ae

N. R. Al Mawali et al. (eds.), *Fourth Industrial Revolution and Business Dynamics*, https://doi.org/10.1007/978-981-16-3250-1_4

55

et al., 2016; Palumbo et al., 2017; Scholl et al., 2014). This trend was accompanied by a growth in studies on PE along the therapeutic continuum (Chang et al., 2012; Nafradi et al., 2018), resulting in many conceptualizations and measurements to operationalize this notion (Cerezo et al., 2016; Fumagalli et al., 2015). PE is defined as the perceived ability of the patient to self-manage own health by getting involved in decisions and assuming responsibility for choices affecting personal health (Schulz & Nakamoto, 2013).

Recognizing the challenges of applying the multidimensional PE concept in empirical settings, scholars synthesized extant knowledge on PE and related dimensions to generate a unified measurement scale (Barr et al., 2015). Although much research was conducted on PE in Western nations, little is known about PE initiatives in the emerging market context of the United Arab Emirates (UAE). The UAE government pursues continuous improvement across an integrated healthcare system, including institutional and service quality, resource usage and cost control, and actual health outcomes for its population (Bodolica et al., 2018; Koornneef et al., 2017; Spraggon & Bodolica, 2014). To achieve this strategic priority, while adapting to the needs of a changing demographic profile and accounting for the surge of several chronic diseases, the UAE embarked on a program of medical reforms and renovations (Koornneef et al., 2017).

Determined to secure a leading position in international rankings, the country aspires to develop a world-class healthcare system and transform itself into a sought-after medical hub on the global arena (Federal Competitiveness and Statistics Authority, 2019; Legatum Institute, 2017; UAE Government, 2018a; World Economic Forum, 2018). The notable progress that has been achieved in the UAE healthcare sector has been well documented in academic and practitioner publications (Abuhejleh et al., 2016; Gachiri, 2017; Weber et al., 2017). Yet, the extent to which members of the public have been empowered to take control of their health and participate in the design, delivery, and governance of healthcare remains unclear. No systematic effort has been deployed to analyze the UAE-based evidence on PE, and we bridge this gap via a two-phased literature review. This chapter offers a critical assessment of the current state of empirical PE research and related constructs in the cultural and regulatory framework of the UAE (Bodolica et al., 2020).

2 METHODOLOGY

As illustrated in Fig. 1, all the review procedures were conducted in accordance with Preferred Reporting Items for Systematic Reviews and Meta-Analyses (PRISMA) guidelines (Al Slamah et al., 2017; Barr et al., 2015; Koornneef et al., 2017). PRISMA guidelines suggest reporting the various steps in the process of records' identification, screening, eligibility, and inclusion, with a clear explanation of reasons for records' exclusion and criteria for inclusion. The purpose of the first phase of review was to clarify the notion of PE and related dimensions. A thorough database search was made to detect refereed scholarly articles published in the 2013–2018 period that reviewed PE research. We focused deliberately on recent review articles as they offer a holistic analysis of the diversified PE-related literature cumulated over the years. Our search was made using ProQuest Health and Medical Complete, PubMed Central, and PsychInfo databases. To generate review-only PE articles, we employed the keyword technique searching for 'PE' and 'review' words simultaneously in papers' title/abstract. After screening the generated records, we removed duplicates and excluded non-reviews and non-PE papers. Five articles were identified by holding discussions with colleagues, but only two were retained as the other three were conceptual papers.

The assessment of articles' full-texts for securing compliance with eligibility criteria was performed by two peers in parallel to increase confidence and eliminate bias (Bodolica & Spraggon, 2018). The eligibility considerations were as follows: articles had to review PE and related literature; be published in peer-reviewed English-language journals; and appear in press from 2013 onward. Empirical and conceptual articles, technical papers/doctoral theses, books/reports, policy briefings, and non-refereed contributions were excluded. The individual results of each peer were compared and one discrepancy emerged regarding a review on patient involvement (Snyder & Engstrom, 2016). This discrepancy was addressed in a meeting, where the decision to drop this paper was made due to its lack of conceptual PE focus.

A final check for sample comprehensiveness was performed using the ancestry approach of articles' identification (Bodolica & Spraggon, 2009, 2015b). It allows examining the reference lists of the most recent articles to determine whether new entries could be generated that were missed through database searching. We screened the references of the six retained articles published in 2016–2017 in search for non-identified

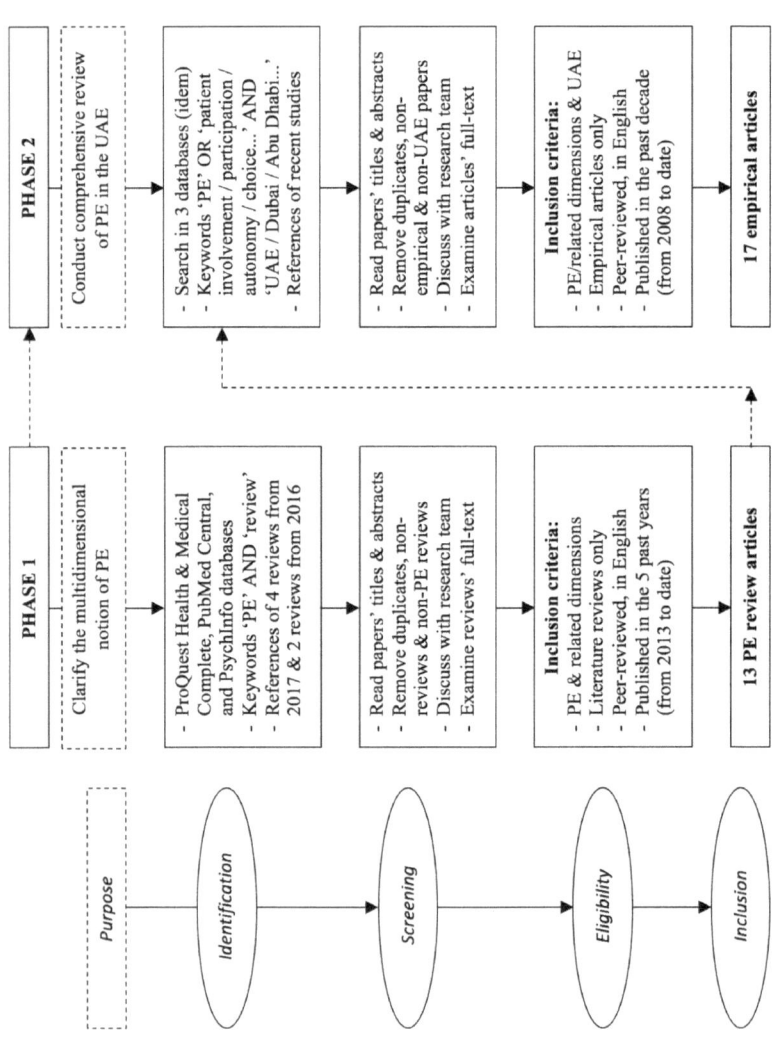

Fig. 1   Methods of the two-phased systematic review of the literature on PE in the UAE

reviews. Since this procedure did not yield additional eligible articles, our final sample of 13 PE-related reviews remained unchanged.

These review articles were examined to derive an accurate definition and understanding of PE and associated dimensions. Islamic scholars highlighted ethical concerns regarding the relevance/timeliness of the principle of individual autonomy and right to self-determination in local settings on the basis of historic, religious, and sociocultural considerations of the Muslim population (Gachiri, 2017; Malek et al., 2018). These concerns could have delayed the widespread acceptance of PE in the UAE, affecting the amount of scholarly production on this topic. Since by focusing on PE exclusively, without considering closely related terms, we could have missed relevant literature, we kept our search as open/comprehensive as possible.

Our analysis revealed that patient autonomy/self-determination, patient choice/voice, patient engagement/participation, patient involvement/activism, patients' rights, self-care/self-management, coping with disease, patient knowledge/information empowerment, and shared decision-making are used interchangeably to infer PE. These terms, separated by the Boolean operator 'or', were entered as keywords when performing the database search. We used the operator 'and' to limit the search to only UAE-based PE studies. To capture locally oriented articles, we alternated between the country name and its three-letter acronym, and employed the names of Abu Dhabi, Dubai, and Sharjah emirates. Since PE topicality is a recent occurrence in the Arab region, we focused on articles published over the past decade to generate an updated account of PE in the UAE.

To complete the second phase of our survey, we followed the procedures delineated above. To decide about the retention of papers, we defined a slightly different set of eligibility criteria. The following conditions for inclusion should have been met: empirical articles on PE and associated constructs in the UAE, published in refereed English-language journals from 2008 onward. Non-academic literature, conceptual papers, and empirical studies that offered an aggregated analysis of PE-related issues in several Arab states were excluded. This was the case of a breast cancer control strategies' study in four geo-regions, where PE emerged as a constitutive dimension of 'promoting advocacy' theme (Bridges et al., 2011). This article was excluded because its findings were presented indiscriminately for a group of 12 Arab nations, without separating the effects for each included country.

The full-texts of 75 generated articles were thoroughly examined to verify their eligibility. Since most of them did not pertain to any dimension of PE, only 16 papers were retained. By screening the references of a recently published survey on diabetes self-management in the Gulf (Al Slamah et al., 2017), we identified one pertinent entry. Each researcher from our team analyzed the contents of the retrieved studies and confirmed their inclusion in the review. The second phase of the process yielded a sample of 17 empirical articles on UAE-based PE issues.

3   RESULTS

3.1   PE-Related Review Articles

Two themes emerged from the content analysis of 13 PE reviews: 'conceptual clarification' and 'contextual embeddedness' (see Table 1). Four first-theme articles are dedicated exclusively to elucidating the notion of PE and associated dimensions. Each survey advances a different conceptual model to map PE-related constructs, highlighting the difficulty of making sense of this complex literature. While some authors focus on PE indicators and behaviors (Bravo et al., 2015), others uncover the antecedents, attributes, and consequences of PE (Castro et al., 2016). The concept is interpreted through the lens of an enabling process that leads to various outcomes (Cerezo et al., 2016), or presented as a combination of ability, motivation, and power (Fumagalli et al., 2015).

Among the most cited PE elements are patient's medical knowledge, coping skills, health literacy/education, information-seeking behavior, gaining control, sense of meaning, shared decision-making, self-care/self-management, and self-efficacy. PE is mapped in relationship with patient activation, enablement, engagement, involvement, participation (Fumagalli et al., 2015), and patient-centeredness, indicating that PE is a broad concept that embraces multiple components (Castro et al., 2016). To improve clinical outcomes and quality of life, PE interventions should be integrated into medical practice through jointly deployed efforts of patients, physicians, and healthcare organizations to induce a long-lasting behavioral change (Bravo et al., 2015; Cerezo et al., 2016).

Nine 'contextual embeddedness' articles conceptualize PE in relation to 'healthcare system', 'healthcare governance', 'information technology' (IT), and 'therapeutic continuum'. Citizenship, culture, and tradition of voluntary action play a critical role in shaping PE at the national

Table 1 Recent literature review articles related to the concept of PE

Reference	Context	Emphasis	PE elements and associated dimensions	Findings and contribution
Theme 1—Conceptual clarification (4 articles)				
Bravo et al. (2015)	A-contextual (concept-focused)	conceptual model of PE (at patient, professional and system level)	indicators (self-efficacy, knowledge, skills, personal control, health literacy, sense of meaning; feeling respected); behaviors (self-management; shared decision making; take part in groups, use Internet to search/share information)	PE is a state ranging from low to high level; responsibilities of patients, providers and healthcare system to use PE interventions to enhance clinical outcomes
Cerezo et al. (2016)	A-contextual (concept-focused)	Analysis of dimensions and measures of PE concept	PE as enabling process (knowledge acquisition, coping skills); PE as outcome (participation in decision making, gaining control); other PE dimensions (self-care, capacity building, trust, motivation, sense of meaning, positive attitude)	Challenge of incorporating PE in healthcare practice through motivational strategies to induce behavioral change

(continued)

Table 1 (continued)

Reference	Context	Emphasis	PE elements and associated dimensions	Findings and contribution
Fumagalli et al. (2015)	A-contextual (concept-focused)	Conceptual map for PE and 5 related concepts	PE as a process, emergent state, and behavior (participation and involvement); PE as a combination of ability (enablement), motivation (engagement) and power (activation)	PE mapped in close relationship with patient activation, enablement, engagement, involvement, and participation
Castro et al. (2016)	A-contextual (concept-focused)	Conceptual analysis of PE, patient participation and patient-centeredness	antecedents (patient education, knowledge, control, participation); attributes (enablement, activation); consequences (self-efficacy, control, self-management); is a much broader concept than patient participation and patient-centeredness	process model in healthcare to improve quality of care/life: strategy of patient participation facilitates patient-centeredness, which leads to PE
Theme 2—Contextual embeddedness (9 articles):				
Topic 2.1—PE and national healthcare system				
Boudioni et al. (2017a)	National healthcare system	Role of citizenship, culture, voluntary community org. in PE in Greece and England	Patient/social/community participation; public involvement; patient informed choice and voice; patients' rights; expert patients; patient ability to control own care	PE shaped by stronger (weaker) citizenship and longer (shorter) tradition of voluntary action in England (Greece)

Reference	Context	Emphasis	PE elements and associated dimensions	Findings and contribution
Boudioni et al. (2017b)	Healthcare policies	Comparison of national policies, systems, structures of PE in Greece and England	Legislation-driven; patients' rights (access to healthcare, quality of care, approval of treatment, respect, consent, confidentiality, information, informed choice, involvement in own healthcare, right of redress); patient-focused services	Policies emphasize: patient-centered services, public involvement and PE, in England; patient rights, responsibilities, and quality of services, in Greece

Topic 2.2—PE and healthcare governance

Reference	Context	Emphasis	PE elements and associated dimensions	Findings and contribution
Bodolica and Spraggon (2014)	Healthcare governance	Divide between macro and micro governance	PE (patient choice, autonomy, medical literacy) as a component of micro-level governance (in the patient-physician relationship)	Advocate the integration of macro and micro governance devices in healthcare settings
Tofan et al. (2013b)	Relational governance	PE as governance mechanism in physician–patient relationship	PE as distrust-based governance tool (patient autonomy, assertive control, information empowerment, choice, involvement, decision-making authority, eHealth, system distrust, use of Internet for information)	Conceptual framework integrating both trust-(doctor-focused) and distrust-based (patient-led) governance

(continued)

Table 1 (continued)

Reference	Context	Emphasis	PE elements and associated dimensions	Findings and contribution
Topic 2.3 —PE and information technology				
Risling et al. (2017)	Electronic health	analysis of PE construct; relationship between PE and eHealth portal usage	involvement in decisions; ability to find mistakes; preparedness; personal control; understanding of provider instructions; patient engagement (self-control, self-management, self-efficacy) and activation (skill, knowledge, confidence for self-care); use of eHealth tech	Huge variety in conceptual operationalization of PE; need to attain definitional consensus and standardized measure of PE to assess its association with the uptake of eHealth solutions
Groen et al. (2015)	Information technology (IT)	conceptual components of PE; contribution of IT to empowerment of cancer survivors	being autonomous and respected; having knowledge; having psychosocial and behavioral skills; perceiving support from community, family and friends; perceiving oneself to be useful	IT (educational, electronic, patient-to-patient, portal services and multicomponent) contributes to PE by enhancing autonomy, knowledge and skills
Calvillo et al. (2013)	Information technology	analysis of how various IT tools contribute to PE	Patient education; health literacy; remote access to health services; access control; self-care; patient as information source; decision making; privacy and confidentiality	IT contributes to PE (more proactive behavior) and allows citizens to act as information providers

Reference	Context	Emphasis	PE elements and associated dimensions	Findings and contribution
Topic 2.4—PE and therapeutic continuum				
te Boveldt et al. (2014)	Cancer pain management	Conceptual model and analysis of PE in controlling pain	Self-efficacy; increasing abilities; locus of control; active coping; participation in decision making; resources (induced by caregiver, skills)	Focus on pain treatment given by clinician, involvement of patient and interaction of both
Nafradi et al. (2017)	Medication adherence	Effect of 2 PE dimensions on adherence	PE components: internal health locus of control (belief of being in control of own health) and self-efficacy (disease management, general)	PE promotes adherence; need for joint empowerment (patients who share control with doctors)

level (Boudioni et al., 2017a), giving rise to healthcare reforms and PE policies and systems (Boudioni et al., 2017b). PE is viewed as a distrust-based (patient-driven) governance attribute in the physician–patient relationship (Tofan et al., 2013b) and a micro-level governance component (Bodolica & Spraggon, 2014). Acknowledging the macro–micro divide in healthcare governance, scholars promote the integration of macro-level (policymaking) and micro-level (doctor–patient interaction) governance initiatives in medical settings. IT-based studies conclude that patients' utilization of e-health tools contributes to their empowerment (Risling et al., 2017) through enhanced autonomy, knowledge, proactive behavior, and self-management (Groen et al., 2015), allowing people to act as seekers and suppliers of medical information (Calvillo et al., 2013). To achieve better outcomes in cancer pain management (te Boveldt et al., 2014) and medication adherence (Nafradi et al., 2017), 'therapeutic continuum' researchers advocate the principle of shared control and joint empowerment of patients and physicians.

PE is a complex multidimensional construct, which: incorporates many tightly intertwined elements (autonomy, self-efficacy, health literacy, information search/use, personal control, coping with illness); is relevant at multiple levels (micro, meso, macro); is analyzed from the standpoint of several stakeholders (patient, clinician, healthcare system); is interpreted in many ways (process, state, behavior, feeling, intervention, outcome); emerges at different stages (antecedents, attributes, consequences); is seen as an intrapersonal disposition (patient power/control) or relational concept (power in clinician–patient relationship); is linked to many disciplines (IT, governance, public policy/administration); oscillates along a continuum (low to high); and is influenced by various moderators (culture, citizenship, legislation, socio-economic conditions, professional goals, health status, education).

3.2   PE-Related Issues in the UAE

Our analysis suggests that most of the 17 UAE-based PE inquiries represent the outcome of repeated efforts of the same researcher teams from local medical colleges/institutions (Table 2). The majority employs cross-sectional surveys and statistical analyses, with only three articles making use of qualitative methodologies to analyze specific case/interview data (Hasan et al., 2016; Laurance et al., 2014; Sulaiman et al., 2009). While adult patients represent the common study subjects, one paper

Table 2 Reviewed empirical studies related to PE in the UAE healthcare settings

Reference	Method	Subjects	Field	PE/related constructs use	Main findings and implications
Theme 1—Chronic disease care (11 studies):					
Topic 1.1—General inquiries					
Hashim et al. (2013)	Cross-sectional survey, regression	38 nurses and physicians attending a workshop	Chronic disease care	n/a / patient self-management support services' usage by clinicians	Clinicians' non-use of proactive patient self-management tools; outreach programs and patient education needed to improve self-care and make chronic disease care systematic (not episodic)

(continued)

Table 2 (continued)

Reference	Method	Subjects	Field	PE/related constructs use	Main findings and implications
Sayiner et al. (2012)	Cross-sectional survey, comparison	27 subjects (out of 1392 across 11 MENA states)	Chronic obstructive pulmonary disease	n/a / knowledge/being informed about disease, information-seeking behavior from various sources	feeling of being informed about respiratory condition is suboptimal (higher in the UAE than MENA); information obtained from doctors, TV and Internet; need for more patient education

Reference	Method	Subjects	Field	PE/related constructs use	Main findings and implications
Topic 1.2—Diabetes management					
Baynouna et al.(2014)	Surveys, regressions	442 patients, 7 centers in Al Ain	Hypertension, diabetes mellitus	PE as perception of being knowledgeable; ability for self-management	Behavior assessment needed to design effective interventions to increase PE and adherence to healthy lifestyle behavior (via self-management)
Hashim et al. (2016)	Cross-sectional survey, regression	165 patients, 2 clinics in Al Ain	Type 2 diabetes mellitus	n/a / disease-related knowledge of patients, self-management	patients' knowledge about diabetes remained low over the 2001–2014 period; education efforts need to focus on behavioral strategies to enable and encourage patients to adopt self-care
Abdulekarem and Sackville (2009)	before-after study, 24 months	59 patients, 3 pharmacies in Sharjah	Type 2 diabetes mellitus	n/a / self-care and self-management (achieved via information reminders sent through pharmacists)	poor disease knowledge, diet and exercise; information programs improve self-management; continuous long-term information/education initiatives needed to induce behavioral change to adopt self-care

(continued)

Table 2 (continued)

Reference	Method	Subjects	Field	PE/related constructs use	Main findings and implications
Sulaiman et al. (2009)	Qualitative, interviews	41 patients, Sharjah	Diabetes	n/a / patients' disease-related knowledge	Knowledge varied; disease attributed to lifestyle, contextual and cultural factors; need for culturally-sensitive strategies to educate about illness
Al-Maskari et al. (2013)	Cross-sectional survey	575 patients, 2 hospitals in Al Ain	Diabetes mellitus	n/a / patient self-management of their chronic disease	Low patient awareness; poor knowledge/skills to self-manage the condition; awareness programs critical to improve coping, adherence and self-care
Topic 1.3—Diabetic complications					
Al-Kaabi et al. (2008)	Cross-sectional study	409 patients, clinics in Al Ain	Diabetes and dietary practice	n/a / self-monitoring or self-management of disease	Poor self-monitoring and dietary practice; patient-tailored dietary counselling needed to empower patients to self-manage their chronic disease

Reference	Method	Subjects	Field	PE/related constructs use	Main findings and implications
Al-Kaabi et al. (2009)	Cross-sectional survey	390 patients, 6 clinics in Al Ain	Diabetes and physical activity	n/a / self-monitoring or self-management of disease	Low level of self-monitoring and physical activity; patient-tailored counselling needed to empower patients to self-manage their chronic disease
Al-Kaabi et al. (2015)	Experimental design, survey	221 illiterate patients in Al Ain	Diabetic foot problems	n/a / illiteracy of patients as predictor of poor foot-related self-care	Illiteracy induces poor knowledge of diabetes and its foot complications; education programs for illiterate patients needed to enhance self-care
Sulaiman et al. (2010)	Cross-sectional survey	347 patients, clinics in Sharjah	diabetes, depression, anxiety	n/a / patient self-care (as correlate of depression)	depressed diabetic patients have poor self-care and adherence; need for self-management initiatives to improve coping with chronic illness

(continued)

Table 2 (continued)

Reference	Method	Subjects	Field	PE/related constructs use	Main findings and implications
Theme 2—Self-medication with drugs (3 studies)					
Shehnaz et al. (2013)	Cross sectional survey	324 expatriate students, 4 schools	Self-medication with drugs	n/a / self-care attitude or autonomous health behavior	High prevalence of self-medication as evidence of taking responsibility for own health but also risk of misuse; education programs needed for making the transition to self-care successful
Shehnaz et al. (2014)	Cross-sectional survey	324 expatriate adolescent students	Self-medication with drugs	n/a / drug knowledge; information-seeking behavior from various sources	low drug knowledge scores, high inclination for self-medication; parents, pharmacists and media as sources of information; need of education campaigns

Reference	Method	Subjects	Field	PE/related constructs use	Main findings and implications
Hasan et al. (2016)	Qualitative, content analysis	30 subjects, purposive sample	self-medication with drugs	n/a / self-medication as an aspect of self-care and sources of drugs' information	Self-medication is common; pharmacists and family as main source of information; pharmacist plays a key role in patient education about self-care
Theme 3 —Non-therapeutic interventions (3 studies)					
Laurance et al. (2014)	Case study, project	21 college students, community	Genetic disease screening	n/a / patient and local community engagement (to educate and spread awareness)	Project benefits: law for premarital screening (due to high consanguinity); higher awareness and life expectancy, lower incidence of disease and cost, better health; need to address cultural sensitivities and build partnerships across levels

(continued)

Table 2 (continued)

Reference	Method	Subjects	Field	PE/related constructs use	Main findings and implications
McLean et al. (2010)	Interviews, clinical scenarios, regressions	218 female Emiratis (Muslims) in Al Ain clinics	Obstetrics, stomach, face, child scenarios	n/a / patient involvement (consent) in students' medical education (for examination by a student)	refusal of cross-gender examination (obstetrics and stomach); need to account for religious and cultural issues; remind patients of their religious duty to contribute towards doctors' training
Al-Yateem and Rossiter (2017)	Cross-sectional survey	300 adolescents, 4 schools in Sharjah	Nutrition and dietary habits	n/a / knowledge of healthy nutrition and diet	adolescents' lack of knowledge of healthy eating and nutrition; to reduce risk of obesity, need to design multifaceted education programs to increase knowledge of healthy nutrition

uses clinical professionals (Hashim et al., 2013) and four employ college students (Al-Yateem & Rossiter, 2017; Laurance et al., 2014), of which two focus on expatriate adolescents (Shehnaz et al., 2013, 2014). Only one investigation by Baynouna et al. (2014) has explicitly mentioned PE (as a perception of being knowledgeable), with most articles referring to various PE dimensions. The two closely related PE constructs that were tackled in UAE studies are patient involvement (McLean et al., 2010) and local community engagement (Laurance et al., 2014). Eleven papers concentrate on patient capacity for self-care/self-management/self-monitoring, with patient disease-related knowledge and information-seeking behavior being examined on six and three occasions, respectively.

We identified three themes on PE-related issues in the UAE: 'chronic disease care', 'self-medication with drugs', and 'non-therapeutic interventions'. The contents of 11 first-theme articles point to three topics: 'general inquiries', 'diabetes management', and 'diabetic complications'. Since public awareness about chronic conditions is suboptimal, the adoption of patient education programs and system interventions is recommended to depart from episodic to more systematic chronic disease care (Hashim et al., 2013; Sayiner et al., 2012). The five 'diabetes management' studies suggest that educational initiatives in the UAE should be deployed continuously to induce sustainable behavioral change for patient espousal of self-care attitudes (Abdulelkarem & Sackville, 2009; Hashim et al., 2016). Since diabetic complications are associated with poor dietary practice, foot problems, low physical activity, depression and anxiety, diabetes counseling should be tailored to patients' needs to improve their coping with chronic illness (Al-Kaabi et al., 2009, 2015).

The second theme includes three studies on self-medication, treated as an aspect of patient self-care. Self-medication, which refers to situations when people administer drugs to treat self-recognized symptoms/sicknesses without professional consultation, represents a means for empowering patients to take control of their health (Abasaeed et al., 2009). Yet, the high inclination for self-medication by UAE adults and adolescents (Hasan et al., 2016) is accompanied by low levels of public knowledge and understanding of medicines and antibiotics (Shehnaz et al., 2014). The practice of self-medication raises ethical concerns and risks of drugs' misuse, indicating that the population might not be well-equipped to take higher responsibility for personal health. To make the

transition to self-care a successful undertaking in the UAE, patient education programs are needed with the active participation of clinicians,
pharmacists, parents, media, and other stakeholders (Hasan et al., 2016;
Shehnaz et al., 2013).

The three 'non-therapeutic interventions' articles address wellbeing
issues that consider UAE's cultural/religious specificities and cut across
many stakeholder groups (adolescents, female Muslims, young Emirati
couples). Given the prevalence of obesity in the UAE, researchers
promote multifaceted educational interventions to enhance adolescents'
knowledge of healthy nutrition and encourage adopting beneficial dietary
habits (Al-Yateem & Rossiter, 2017). Since consanguineous marriages are
practiced by Muslim couples, local community engagement is critical for
overcoming culturally induced resistance and spreading awareness about
genetic disease screening to drive the implementation of a premarital
screening law (Laurance et al., 2014). In a study of patient involvement in
students' medical education, Emirati women with gynecological problems
refused to submit to cross-gender examinations (McLean et al., 2010).
While addressing cultural sensitivities is important, female patients need to
be reminded of their social/religious duty to contribute toward doctors'
training in the country.

4   DISCUSSION

By juxtaposing themes and topics from PE reviews and UAE-based
studies, we delineate three opportunities for future inquiry and policy
intervention on PE in the UAE (Fig. 2).

4.1   Opportunity (1)—Consolidate

Most UAE studies connect PE with therapeutic continuum aspects of
chronic disease care and self-medication. This provides opportunities for
consolidation by conducting confirmatory research on larger samples
across different Emirates to build a foundation for making generaliza-
tions. Unsurprisingly, most sampled studies relate to diabetes and its
complications, since 20% of the total UAE population is battling this
chronic condition (Alhyas et al., 2011). PE plays a critical role in the
successful management of chronic illness, where the onus is on the patient
to embrace the logic of active coping with disease through disciplined

Fig. 2   Opportunities derived from juxtaposing themes/topics from Tables 1 and 2

self-monitoring (Lorig & Holman, 2003; Sayiner et al., 2012; Zimbudzi et al., 2015). Yet, the diabetes-related knowledge, information-seeking behavior, and self-management attitudes of UAE patients remain weak (Al-Kaabi et al., 2008, 2009; Sulaiman et al., 2009, 2010). The UAE Government ambitions to reduce the percentage of diabetes by 2021, but the attainment of this goal depends on the effectiveness of PE programs to encourage optimal levels of patient self-care (UAE Government, 2018b).

In an analysis of health status in the UAE, cardiovascular diseases, injury, cancers, and respiratory disorders were identified as public health priorities to be addressed at the national level (Loney et al., 2013). Further assessments are needed on the contribution of PE strategies to the enhancement of health outcomes of patients with these chronic conditions through higher self-efficacy, pain management (te Boveldt et al., 2014), and medication adherence (Nafradi et al., 2017). Extant studies on self-medication in the UAE focus on its risks and negative health implications due to the gap between patients' state of 'feeling' and 'being' informed about drugs and medical principles (Shehnaz et al., 2013, 2014). Authors noted that the prevalence of antibiotics' self-medication in Abu Dhabi may reflect both the lack of punitive legislation for pharmacies dispensing drugs without prescription and the demographic aspect of the Emirate where its expatriate majority relies on home-country sources of medicines (Abasaeed et al., 2009). Regulatory interventions and educational programs aimed at reducing the incidence of drugs' misuse will help refocusing researchers' attention on the study of beneficial aspects of self-medication as a manifestation of autonomous health behavior.

4.2   Opportunity (2)—Expand

The sporadic and inconsistent use of PE suggests that the topic requires deeper exploration in the UAE. The propensity of decision-makers, clinical professionals, and patient advocates to discuss PE is the lowest in Arab countries, compared to the Canada–Australia cluster and even Latin American and Asian nations (Bridges et al., 2011). Given the embryonic stage of development and scarcity of relevant UAE-based studies, many opportunities for expanding inquiry exist by combining insights from 'conceptual clarification' and 'non-therapeutic interventions' themes. To design viable interventions (Laurance et al., 2014; McLean et al., 2010), we recommend delving deeper into PE and its contextual application by considering social, cultural, and religious characteristics of the UAE (Bodolica & Spraggon, 2015a).

The difficulty of achieving definitional consensus on PE (Castro et al., 2016; Cerezo et al., 2016) is acknowledged due to the variability of national settings where the concept is employed, inferring asymmetric levels of literacy and access to information, concerns about digital divide and availability of Internet, and confidentiality issues (Risling et al., 2017). We call for contextualizing conceptual clarity efforts through a measurement scale that would allow operationalizing PE within the locally relevant value sets of the UAE. A hindrance for PE interventions may be the low literacy rates among older Emiratis and unskilled expatriate workers who lack formal education (Koornneef et al., 2017). Developing a reliable health literacy screening instrument that would be culturally specific to target idiosyncrasies of the UAE socio-economic fabric represents a step forward (Nair et al., 2016). There is more scope for expanding research/practice on the effectiveness of PE methods directed to the youth to inculcate a mentality of healthy nutrition (Al-Yateem & Rossiter, 2017), as obesity reduction among children represents another target of the 2021 UAE National Agenda (UAE Government, 2018b).

Although PE is gaining traction in global markets, some cultures might not be ready to embrace the trend toward increased autonomy and self-determination. In Muslim countries, patients may prefer to rely on professionals' expert opinions or concede their individual decision-making power to their (male) family members (Malek et al., 2018). From the perspective of Islam, while people enjoy the freedom of self-governance, the principles of beneficence and non-maleficence are given priority in medical decision-making, especially when patients are uninformed and possess limited understanding of their disease (Rathor et al., 2011). To decide about a treatment while accounting for patients' lack of competence, Islamic teachings reserve a central place to physicians, due to their professional and religious duty to do good and ward off harm (Elbarazi et al., 2017). Faith-based, community-empowered, and family centered participatory approaches may be appealing to the Muslim UAE majority, where Islamic principles and religious obligations form part of daily life (Jamali et al., 2018). The role that community leaders, places of worship, and the family can play in spreading awareness about healthy lifestyles and empowering the UAE population to self-manage their health is worthy of further exploration (Gachiri, 2017).

4.3   Opportunity (3)—Initiate

Our analysis unveils a major decoupling between PE topics in review articles and those examined in the UAE context. The dearth of UAE-embedded inquiries on PE in relation to 'national healthcare system', 'healthcare governance', and 'IT' provides opportunities for initiating research, reforms, and practice in these areas. Since PE is viewed as a fundamental pillar in the development of a sustainable healthcare ecosystem (Palumbo et al., 2017), decision-makers ought to craft initiatives that would boost patient participation across levels of an integrated healthcare system. The future makeup of medical practice and the implementation of a patient-centered approach to care depend on PE strategies that are formulated today (Boudioni et al., 2017a, 2017b). UAE legislators and practitioners should revisit institutional and clinical arrangements in health service provision to offer more room for residents to get involved in the design and delivery of care and play a heightened role in healthcare governance (Bodolica & Spraggon, 2014).

A greater sense of health ownership could be developed through educational policies and supporting infrastructures that would empower patients in medical encounters. People should engage in health-related initiatives in their community and voice their opinions regarding national priorities for policymaking. Scholars could assess the effectiveness of patient-directed interventions in transforming people into value creators and active participants in healthcare markets (Dent & Pahor, 2015).

Federal public health frameworks should be revised periodically to secure compliance with international best practices and alignment with dominant health concerns. The diversity of the UAE population (age/gender distribution, educational/economic backgrounds, social/cultural characteristics) poses challenges for the design of adequate public health reforms (Loney et al., 2013). PE education and intervention methods should be culturally sensitive, embedded in the nation's social fabric (Bodolica et al., 2015), and tailored to the needs of a specific group (Laurance et al., 2014).

Digital era technologies represent valuable platforms for drawing on citizens' insights to transform public institutions and policymaking (Risling et al., 2017). Although health websites allow empowering Saudi Arabia patients (Househ et al., 2013, 2014), studies on how the adoption of electronic health systems drive PE in the UAE are lacking. Only one inquiry examined clinicians' viewpoints about e-health development challenges in the UAE compared to other Arab states, but no associations were made with PE-related consequences (Uluc & Ferman, 2016). In technology-savvy nations, a tighter integration of IT into the medical sector may offer benefits in terms of health outcomes and general well-being (Gachiri, 2017). In the 2016 Global IT Report, UAE is ranked 26th worldwide and 1st in Middle East on the Networked Readiness Index, unveiling a high level of government usage and social impact of IT (World Economic Forum, 2016). Considering the ever-expanding role of digital technologies and electronic portals in the UAE medical landscape, more research is needed on how e-health contributes to PE.

UAE residents are becoming increasingly active on social media, rely on cellphone apps to make decisions, participate in online forums and support groups, and employ media channels to access health-related data (Gachiri, 2017). While social media usage for health information is an indicator of PE, this technology is associated with data inaccuracies, limited usability, misinformation, and privacy/security issues (Househ et al., 2014). If information-seeking behavior is deployed as a tool for 'empowering' rather than 'misleading' patients (Dahl & Eagle, 2016), healthcare organizations have to ensure the readability of data available online to improve people's health literacy. Clinicians should fulfill their moral obligation of facilitating PE by directing patients to health websites that are reliable and trustworthy (Spraggon & Bodolica, 2015). Studies on the role of social, educational, and economic factors in the information-seeking behavior of empowered patients could be insightful for disseminating digitally the medical information that people can comprehend and act upon to solve their health-related concerns.

Acknowledgements Reprinted from Public Health, 169/4, Bodolica V. and Spraggon M., Toward patient-centered care and inclusive healthcare governance: A review of patient empowerment in the UAE, 114-124, Copyright (2019), with permission from Elsevier.

REFERENCES

Abasaeed, A., Vlcek, J., Abuelkhair, M., & Kubena, A. (2009). Self-medication with antibiotics by the community of Abu Dhabi Emirate, United Arab Emirates. *Journal of Infection in Developing Countries, 3*(7), 491–497.

Abdulelkarem, A. R., & Sackville, M. A. (2009). Changes of some health indicators in patients with type 2 diabetes: A prospective study in three community pharmacies in Sharjah, United Arab Emirates. *Libyan Journal of Medicine, 4*(1), 31–36.

Abuhejleh, A., Dulaimi, M., & Ellahham, S. (2016). Using lean management to leverage innovation in healthcare projects: Case study of a public hospital in the UAE. *BMJ Innovations, 2*, 22–32.

Alhyas, L., McKay, A., Balasanthiran, A., & Majeed, A. (2011). Quality of type 2 diabetes management in the states of the Co-operation Council for the Arab states of the Gulf: A systematic review. *PLoS ONE, 6*(8), e22186.

Al-Kaabi, J., Al-Maskari, F., Saadi, H., Afandi, B., Parkar, H., & Nagelkerke, N. (2008). Assessment of dietary practice among diabetic patients in the United Arab Emirates. *Review of Diabetic Studies, 5*(2), 110–115.

Al-Kaabi, J., Al-Maskari, F., Saadi, H., Afandi, B., Parkar, H., & Nagelkerke, N. (2009). Physical activity and reported barriers to activity among type 2 diabetic patients in the United Arab Emirates. *Review of Diabetic Studies, 6*(4), 271–278.

Al-Kaabi, J. M., Al Maskari, F., Cragg, P., Afandi, B., & Souid, A.-K. (2015). Illiteracy and diabetic foot complications. *Primary Care Diabetes, 9*, 465–472.

Al-Maskari, F., El-Sadig, M., Al-Kaabi, J. M., Afandi, B., Nagelkerke, N., & Yeatts, K. B. (2013). Knowledge, attitude and practices of diabetic patients in the United Arab Emirates. *PLoS ONE, 8*(1), e52857.

Al Slamah, T., Nicholl, B. I., Alslail, F. Y., & Melville, C. A. (2017). Self-management of type 2 diabetes in Gulf Cooperation Council countries: A systematic review. *PLoS ONE, 12*(12), e0189160.

Al-Yateem, N., & Rossiter, R. (2017). Nutritional knowledge and habits of adolescents aged 9 to 13 years in Sharjah, United Arab Emirates: A cross-sectional study. *East Mediterranean Health Journal, 23*(8), 551–558.

Barr, P. J., Scholl, I., Bravo, P., Faber, M. J., Elwyn, G., & McAllister, M. (2015). Assessment of patient empowerment—A systematic review of measures. *PLoS ONE, 10*(5), 1–24.

Baynouna, L. M., Neglekerke, N. J. D., Ali, H. E., ZeinAlDeen, S. M., & Al Ameri, T. A. (2014). Audit of healthy lifestyle behaviors among patients with diabetes and hypertension attending ambulatory health care services in the United Arab Emirates. *Global Health Promotion, 21*(4), 44–51.

Bodolica, V., Spraggon, M., & Saleh, N. (2020). Innovative leadership in leisure and entertainment industry: The case of the UAE as a global tourism hub. *International Journal of Islamic and Middle Eastern Finance and Management, 13*(2), 323–337.

Bodolica, V., & Spraggon, M. (2018). An end-to-end process of writing and publishing influential literature review articles: Do's and don'ts. *Management Decision, 56*(11), 2472–2486.

Bodolica, V., & Spraggon, M. (2014). Clinical governance infrastructures and relational mechanisms of control in healthcare organizations. *Journal of Health Management, 16*(2), 183–198.

Bodolica, V., & Spraggon, M. (2015a). Life on heels and making deals: A narrative approach to female entrepreneurial experiences in the UAE. *Management Decision, 53*(5), 984–1004.

Bodolica, V., & Spraggon, M. (2015b). *Mergers and acquisitions and executive compensation.* Routledge.

Bodolica, V., & Spraggon, M. (2009). Merger and acquisition transactions and executive compensation: A review of the empirical evidence. *Academy of Management Annals, 3*(1), 109–181.

Bodolica, V., Spraggon, M., & Shahid, A. (2018). Strategic adaptation to environmental jolts: An analysis of corporate resilience in the property development sector in Dubai. *Middle East Journal of Management, 5*(1), 1–20.

Bodolica, V., Spraggon, M., & Tofan, G. (2016). A structuration framework for bridging the macro-micro divide in healthcare governance. *Health Expectations, 19*(4), 790–804.

Bodolica, V., Spraggon, M., & Zaidi, S. (2015). Boundary management strategies for governing family firms: A UAE-based case study. *Journal of Business Research, 68*(3), 684–693.

Boudioni, M., McLaren, S., & Lister, G. (2017a). The role of citizenship, culture and voluntary community organizations towards patient empowerment in England and Greece. *International Journal of Caring Sciences, 10*(1), 303–312.

Boudioni, M., McLaren, S., & Lister, G. (2017b). A critical analysis of national policies, systems, and structures of patient empowerment in England and Greece. *Patient Preference and Adherence, 11*, 1657–1669.

Bravo, P., Edwards, A., Barr, P. J., Scholl, I., Elwyn, G., & McAllister, M. (2015). Conceptualizing patient empowerment: A mixed methods study. *BMC Health Services Research, 15*, 252.

Bridges, J. F. P., Anderson, B. O., Buzaid, A. C., Jazieh, A. R., Niessen, L. W., Blauvelt, B. M., & Buchanan, D. R. (2011). Identifying important breast cancer control strategies in Asia, Latin America and the Middle East/North Africa. *BMC Health Services Research, 11*, 227.

Calvillo, J., Roman, I., & Roa, L. M. (2013). How technology is empowering patients? *A Literature Review. Health Expectations, 18*(5), 643–652.

Castro, E. A., Regenmortel, T. V., Vanhaecht, K., Sermeus, W., & Hecke, A. V. (2016). Patient empowerment, patient participation and patient-centeredness in hospital care: A concept analysis based on a literature review. *Patient Education and Counseling, 99*, 1923–1939.

Cerezo, P. G., Juve-Udina, M. E., & Delgado-Hito, P. (2016). Concepts and measures of patient empowerment: A comprehensive review. *Revista Da Escola De Enfermagem Da USP, 50*(4), 664–671.

Chang, A. K., Fritschi, C., & Kim, M. J. (2012). Nurse-led empowerment strategies for hypertensive patients with metabolic syndrome. *Contemporary Nurse, 42*, 118–128.

Dahl, S., & Eagle, L. (2016). Empowering or misleading? Online health information provision challenges. *Marketing Intelligence & Planning, 34*(7), 1000–1020.

Dent, M., & Pahor, M. (2015). Patient involvement in Europe—A comparative framework. *Journal of Health Organization and Management, 29*(5), 546–555.

Elbarazi, I., Devlin, N. J., Katsaiti, M.-S., Papadimitropoulos, E. A., Shah, K. K., & Blair, I. (2017). The effect of religion on the perception of health states among adults in the United Arab Emirates: A qualitative study. *BMJ Open, 7*, e016969.

Federal Competitiveness and Statistics Authority. (2019). *Government and private sector health services statistics.* Available at http://fcsa.gov.ae/en-us.

Fumagalli, L. P., Radaelli, G., Lettieri, E., Bertele, P., & Masella, C. (2015). Patient empowerment and its neighbors: Clarifying the boundaries and their mutual relationships. *Health Policy, 119*, 384–394.

Gachiri, W. (2017). A gradient analysis of economic development and chronic diseases. Case of diabetes and life expectancy in United Arab Emirates. *Journal of Applied Business Economics, 19*(1), 35–43.

Groen, W. G., Kuijpers, W., Oldenburg, H. S., Wouters, M. W., Aaronson, N. K., & van Harten, W. H. (2015). Empowerment of cancer survivors through information technology: An integrative review. *Journal of Medical Internet Research, 17*(11), e270.

Hasan, S., Farghadani, G., Al Haideri, S. K., & Fathy, M. A. (2016). Pharma-cist opportunities to improve public self-medicating practices in the UAE. *Pharmacology & Pharmacy, 7*, 459–471.

Hashim, M. J., Mustafa, H., & Ali, H. (2016). Knowledge of diabetes among patients in the United Arab Emirates and trends since 2001: A study using the Michigan Diabetes Knowledge Test. *East Mediterranean Health Journal, 22*(10), 742–748.

Hashim, M. J., Prinsloo, A., & Mirza, D. M. (2013). Quality improvement tools for chronic disease care—More effective processes are less likely to be implemented in developing countries. *International Journal of Health Care Quality Assurance, 26*(1), 14–19.

Househ, M., Alsughayar, A., & Al-Mutairi, M. (2013). Empowering Saudi patients: How do Saudi health websites compare to international health websites? *Studies in Health Technology and Informatics, 183*, 296–301.

Househ, M., Borycki, E., & Kushniruk, A. (2014). Empowering patients through social media: The benefits and challenges. *Health Informatics Journal, 20*(1), 50–58.

Jamali, D., Bodolica, V., & Lapina, Y. (2018). *Corporate governance in Arab countries: Specifics & outlooks*. Virtus Interpress.

Koornneef, E., Robben, P., & Blair, I. (2017). Progress and outcomes of health systems reform in the United Arab Emirates: A systematic review. *BMC Health Services Research, 17*, 672.

Laurance, J., Henderson, S., Howitt, P. J., Matar, M., Al Kuwari, H., Edgman-Levitan, S., & Darzi, A. (2014). Patient engagement: Four case studies that highlight the potential for improved health outcomes and reduced costs. *Health Affairs, 33*(9), 1627–1634.

Legatum Institute. (2017). *The Legatum Prosperity Index*. Available at www.pro perity.com.

Loney, T., Aw, T.-C., Handysides, D. G., Ali, R., Blair, I., Grivna, M., et al. (2013). An analysis of the health status of the United Arab Emirates: The 'Big 4' public health issues. *Global Health Action, 6*, 20100.

Lorig, K. R., & Holman, H. (2003). Self-management education: History, defi-nition, outcomes, and mechanisms. *Annals of Behavioral Medicine, 26*(1), 1–7.

Malek, M. M., Abdul Rahman, N. N., Hasan, M. S., & Abdullah, L. H. (2018). Islamic considerations on the application of patient's autonomy in end-of-life decision. *Journal of Religious Health, 57*(4), 1524–1537.

McLean, M., Al Ahbabi, S., Al Ameri, M., Al Mansoori, M., Al Yahyaei, F., & Bernsen, R. (2010). Muslim women and medical students in the clinical encounter. *Medical Education, 44*, 306–315.

Moretta Tartaglione, A., Cavacece, Y., Cassia, F., & Russo, G. (2018). The excellence of patient-centered healthcare: Investigating the links between empowerment, co-creation and satisfaction. *Total Quality Management Journal, 30*(2), 153–167.

Nafradi, L., Nakamoto, K., Csabai, M., Papp-Zipernovszky, O., & Schulz, P. J. (2018). An empirical test of the Health Empowerment Model: Does patient empowerment moderate the effect of health literacy on health status? *Patient Education and Counseling, 101*, 511–517.

Nafradi, L., Nakamoto, K., & Schulz, P. J. (2017). Is patient empowerment the key to promote adherence? A systematic review of the relationship between self-efficacy, health locus of control and medication adherence. *PLoS ONE, 12*(10), e0186458.

Nair, S. C., Satish, K. P., Sreedharan, J., & Ibrahim, H. (2016). Assessing health literacy in the Eastern and Middle-Eastern cultures. *BMC Public Health, 16*, 831.

Palumbo, R., Cosimato, S., & Tommasetti, A. (2017). Dream or reality? A recipe for sustainable and innovative health care ecosystems. *Total Quality Management Journal, 29*(6), 847–862.

Rathor, M. Y., Rani, M. F., Shah, A. S., Leman, W. I., Akter, S. F., & Omar, A. M. (2011). The principle of autonomy as related to personal decision-making concerning health and research from an 'Islamic viewpoint.' *Journal of Islamic Medical Association in North America, 43*(1), 27–34.

Risling, T., Martinez, J., Young, J., & Thorp-Froslie, N. (2017). Evaluating patient empowerment in association with eHealth technology: Scoping review. *Journal of Medical Internet Research, 19*(9), e329.

Sayiner, A., Alzaabi, A., Obeidatc, N. M., Nejjari, C., Beji, M., & Uzaslan, E. (2012). Attitudes and beliefs about COPD: Data from the BREATHE study. *Respiratory Medicine, 106*(S2), S60–S74.

Scholl, I., Zill, J. M., Harter, M., & Dirmaier, J. (2014). An integrative model of patient-centeredness—A systematic review and concept analysis. *PLoS ONE, 9*(9), e107828.

Schulz, P. J., & Nakamoto, K. (2013). Health literacy and patient empowerment in health communication: The importance of separating conjoined twins. *Patient Education and Counseling, 90*, 4–11.

Shehnaz, S. I., Khan, N., Sreedharan, J., & Arifulla, M. (2014). Drug knowledge of expatriate adolescents in the United Arab Emirates and their attitudes towards self-medication. *International Journal of Adolescent Medicine and Health, 26*(3), 423–431.

Shehnaz, S. I., Khan, N., Sreedharan, J., Issa, K. J., & Arifulla, M. (2013). Self-medication and related health complaints among expatriate high school students in the United Arab Emirates. *Pharmacy Practice, 11*(4), 211–218.

Snyder, H., & Engstrom, J. (2016). The antecedents, forms and consequences of patient involvement: A narrative review of the literature. *International Journal of Nursing Studies, 53*, 351–358.

Spraggon, M., & Bodolica, V. (2014). *Managing organizations in the United Arab Emirates: Dynamic characteristics and key economic developments.* Palgrave Macmillan.

Spraggon, M., & Bodolica, V. (2015). Trust, authentic pride and moral reasoning: A unified framework of relational governance and emotional self-regulation. *Business Ethics: A European Review, 24*(3), 297–314.

Sulaiman, N., Hamdan, A., Al-Bedri, D. A. M., & Young, D. (2009). Diabetes knowledge and attitudes towards prevention and health promotion: Qualitative study in Sharjah, United Arab Emirates. *International Journal of Food Safety, Nutrition and Public Health, 2*(1), 78–88.

Sulaiman, N., Hamdan, A., Tamim, H., Mahmood, D. A., & Young, D. (2010). The prevalence and correlates of depression and anxiety in a sample of diabetic patients in Sharjah, United Arab Emirates. *BMC Family Practice, 11*, 80.

te Boveldt, N., Vernooij-Dassen, M., Leppink, I., Samwel, H., Vissers, K., & Engels, Y. (2014). Patient empowerment in cancer pain management: An integrative literature review. *Psycho-Oncology, 23*, 1203–1211.

Tofan, G., Bodolica, V., & Spraggon, M. (2013). Governance mechanisms in the physician-patient relationship: A literature review and conceptual framework. *Health Expectations, 16*(1), 14–31.

Tofan, G., Spraggon, M., & Bodolica, V. (2013). Agency problems, ethical challenges and governance attributes in different models of physician–patient interaction within the assisted reproduction setting. *Public Health, 127*(6), 597–600.

UAE Government. (2018a). *Good health and wellbeing.* Available at https://government.ae/en/about-the-uae/leaving-no-one-behind/3goodhealthandwellbeing.

UAE Government. (2018b). *Health and fitness.* Available at https://government.ae/en/information-and-services/health-and-fitness.

Uluc, N. C. I., & Ferman, M. (2016). A comparative analysis of user insights for e-health development challenges in Turkey, Kingdom of Saudi Arabia, Egypt and United Arab Emirates. *Journal Management, Marketing and Logistics, 3*(2), 176–189.

Weber, A. S., Turjoman, R., Shaheen, Y., Al Sayyed, F., Hwang, M. J., & Malick, F. (2017). Systematic thematic review of e-health research in the Gulf Cooperation Council (Arabian Gulf): Bahrain, Kuwait, Oman, Qatar, Saudi Arabia and United Arab Emirates. *Journal of Telemedicine and Telecare, 23*(4), 452–459.

World Economic Forum. (2016). *The Global Information Technology Report.* Available at https://www.weforum.org.

World Economic Forum. (2018). *The Global Competitiveness Report 2017–2018.* Available at https://www.weforum.org.

Zimbudzi, E., Lo, C., Misso, M., Ranasinha, S., & Zoungas, S. (2015). Effectiveness of management models for facilitating self-management and patient outcomes in adults with diabetes and chronic kidney disease. *Systems Review, 4,* 81.

Cost–Benefit Analysis and Environmental Impact Assessment of 3D Printing Applications in Building Construction in Oman

Vineet Tirth and Syed Waheedullah Ghori

1 Introduction

Oman is an important country in the Middle East at a key strategic location. The Gross Domestic Product (GDP) of Oman in the year 2018 was 79.29 billion United States Dollars (USD), which registered a growth rate of 12% from the preceding year (Trading Economics, 2019). The GDP of Oman from 2009 to 2018 and its growth rate are shown in Fig. 1. The GDP growth rate in the year 2017 was negative due to the downfall in the oil prices and the instability in the geopolitical situations in the Middle East but it displayed a remarkable transformation, recording a growth rate of 12% in the year 2018. Infrastructure development is a distinct indicator of the GDP of a country.

V. Tirth (✉) · S. W. Ghori
Mechanical Engineering Department, College of Engineering, King Khalid University, Abha, Saudi Arabia
e-mail: vtirth@kku.edu.sa

S. W. Ghori
e-mail: sgori@kku.edu.sa

© The Author(s), under exclusive license to Springer Nature Singapore Pte Ltd. 2021
N. R. Al Mawali et al. (eds.), *Fourth Industrial Revolution and Business Dynamics*, https://doi.org/10.1007/978-981-16-3250-1_5

89

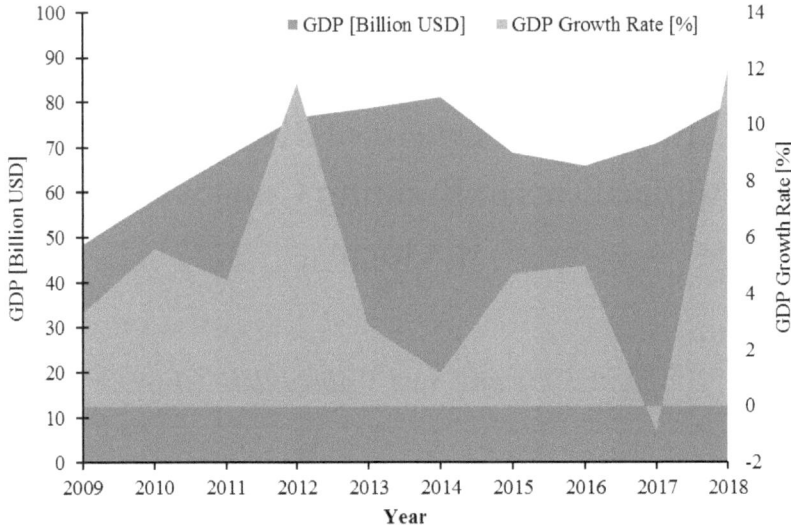

Fig. 1 The GDP of Oman and its growth rate from 2009 to 2018 (*Source* Prepared by the authors based on data source from Trading Economics, 2019)

The total amount spent in USD toward the infrastructure development in Oman from 2009 to 2018 and its percentage contribution to the overall GDP of Oman is shown in Fig. 2. The data has been obtained in Oman Riyals (OMR) from Trading Economics, 2019 and it has been converted to the USD using an online converter (XE Currency Converter, 2019) and plotted. The maximum growth in the infrastructure has been marked in the year 2016, which is a nominal 1.33% of the total GDP (Business Live, 2019; Statista, 2019). The percent contribution of infrastructure development in the total GDP of the top seventeen countries ranked according to the growth of GDP is presented in Fig. 3. The data has been obtained from Statista, 2019 and the plot was developed. The contribution of infrastructure growth in the overall GDP is recorded highest by China (8.3%), followed by India (5.6%), and so on. The maximum contribution of infrastructure in the total GDP for Oman was recorded in the year 2016, which was 1.33% while it dropped to 0.96% in 2018. It is much less than the fast-developing countries indicating that the infrastructure development in Oman is lagging.

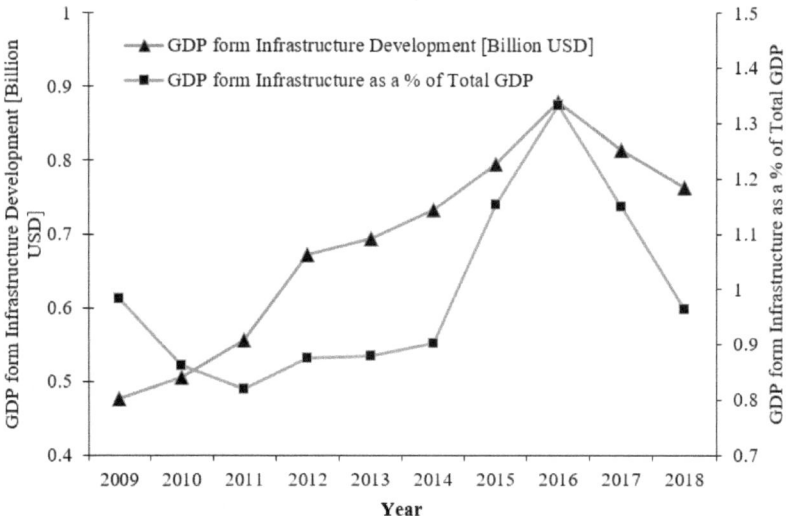

Fig. 2   The GDP of Oman from construction and its percentage contribution to total GDP (*Source* Prepared by the authors based on data source from Trading Economics, 2019)

Based on the human development index (HDI), the countries are categorized as developed and developing, which depends on three major factors such as life expectancy, education, and the gross national income (GNI) per capita (UNDP, 2018).

Figure 4 represents the recent top fifteen developed countries and their HDI for the year 2017. Norway tops the list with HDI of 0.94. Oman has reserved a place in the list of top ten developing countries with its HDI 0.82, shown in Fig. 5. The United States of America is not included in the tally because its HDI is the highest since the 1990s and it maintains the space in the top developed countries (UNDP, 2018).

Infrastructure development is a key component of the HDI. The infrastructure development in Oman is quite less than what is required to reserve its place in the top developing countries and hence it needs special focus. The Vision, 2040 on Oman (Oman Vision 2040, 2019) includes a competitive economy and sustainable living for which, affordable housing and other infrastructure development may play a key role as that in the

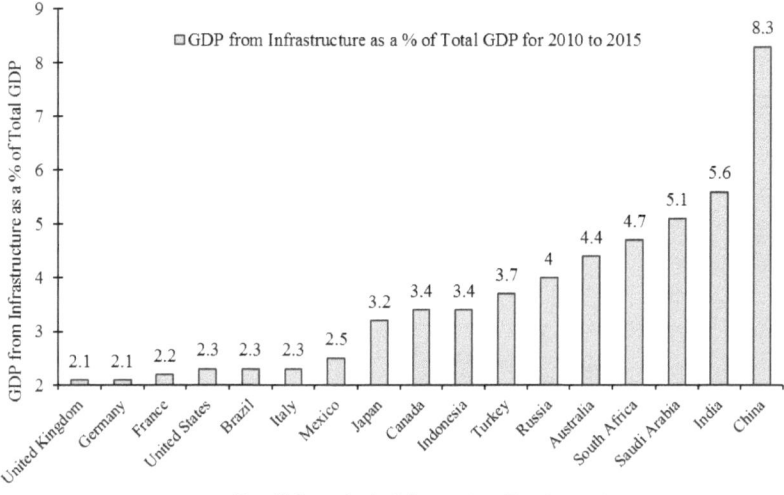

Fig. 3   The percent contribution of infrastructure in total GDP of the top 17 countries (*Source* Prepared by the authors based on data source from Statista, 2019)

top developing country (China). The percent contribution of infrastructure development in the overall GDP needs to be increased and reach around 5% to support its Vision, 2040. The infrastructure development has to be sustainable with the optimum utilization of natural resources. The government of Oman is promoting a private investment of 29% in infrastructure development (Business Live, 2019). The introduction of 3 Dimensional Printing (3DP) in the infrastructure development in Oman may offer one outstanding initiative to achieve its objectives. Furthermore, in the year 2018, 85.9% population of Oman was urban, which is projected to increase to 87% by the year 2020, and 92.5% by the year 2040 (Worldometers, 2019). To meet the requirements of urbanization, quick, and sustainable infrastructure development at a faster pace is required in Oman.

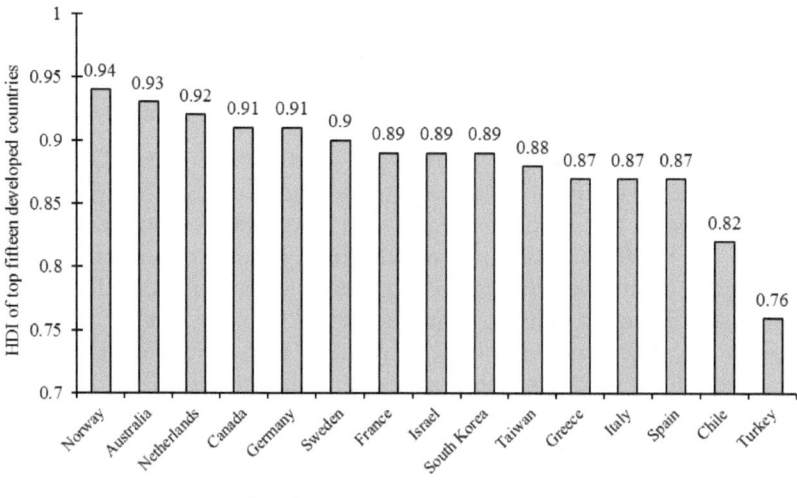

Fig. 4 The HDI of the top 15 developed countries (*Source* Prepared by the authors based on data source from UNDP, 2018)

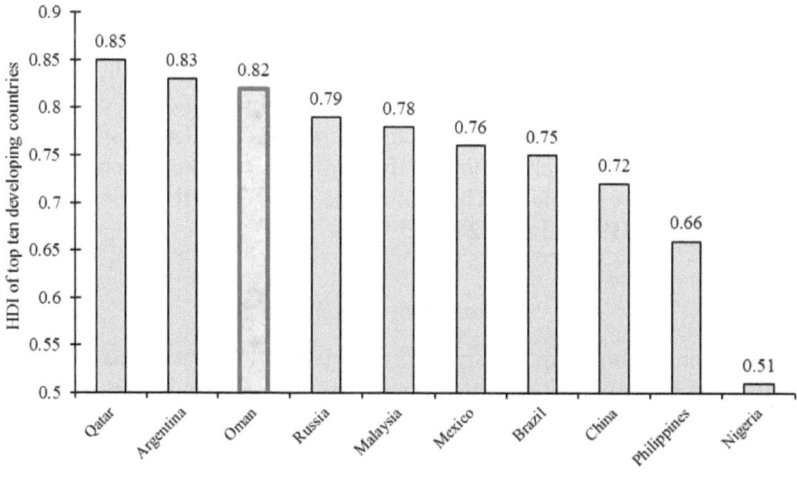

Fig. 5 The HDI of the top 10 developing countries including Oman (*Source* Prepared by the authors based on data source from UNDP, 2018)

2   LITERATURE REVIEW
AND ANALYSIS OF RELATED WORK

2.1   Role of 3DP in Sustainable Infrastructure Development

The conventional building construction methods are not sufficient to meet the demand for infrastructure development of the society. Affordable housing is the need of the day and traditional construction techniques are far from meeting this call. It is of utmost importance to build high-quality infrastructure, economically and rapidly. However, even state-of-the-art building construction technology cannot meet this expectation. The major economic limitations of the traditional building construction methods are labor-intensive and skill depended on activities, which have certain drawbacks such as susceptibility to human errors, transportation cost and interrupted logistic supply, the slow pace of construction, costly materials due to the supply from several vendors, and huge construction wastes. The environmental issues in the construction of the traditional construction methods are huge construction waste generation, emission factors in manufacturing and transportation of the construction materials, energy-intensive and therefore emission-intensive processes, depletion of the natural resources, etc. All these issues lead to a three-fold problem. Firstly, the high cost of materials and labor, delays, and wastage, which has to be borne by the consumers. Secondly, wastage of natural resources in materials and excessive consumption of energy for material transportation and construction activities. The latter issue affects the environment severely. Thirdly, the high cost of construction, slow infrastructure development, which limit the GDP growth of the country (Aimar et al., 2019; Kidwell, 2017).

2.2   Automation in Construction

Automation in building construction started in 1970 (García de Soto et al., 2018) with the introduction of the robotic construction but it was too inferior to the 3DP construction technology because the robot is heavy, it is sensitive to the environment and it has a limited range, size, and mobility. The average permissible load on a roof slab is about 0.6 tons per m^2. The arms of the robot have a limited reach and once it transfers heavy loads of the construction materials, the accidents of toppling over are common (Paoletti & Ceccon, 2018). It is difficult to transfer the construction robot at different levels of the buildings, as the floors are cast. These issues may be conveniently addressed by the application of additive manufacturing in building construction by using 3DP to develop affordable housing. Therefore, 3DP is a viable option for a reliable, versatile, and automated construction system. The core design of the 3DP equipment for building construction consists of a self-propelled 3DP and a material feed system. Such a system can print the modules of the building in parts or as an integrated component such as the concrete blocks for foundation, the walls, columns and the roof slabs, the door and window frames, pavements, etc. The building structure can be printed in an integrated process, or different 3DP modules may be assembled to complete the basic civil structure, quickly and economically. The electrical, plumbing, glasswork, woodworking, and finishing work is to be done by the conventional methods until a more comprehensive multimaterial 3DP technology is commercialized. The hardware used in 3DP construction applications includes an agile, mobile, and lightweight 3DP, which can be easily transported. The 3DP equipment is small and highly stable. It is connected with an automatic construction mix and supply system to maintain a stable flow of the building materials such as fiber concrete, replacing all the complex and dirty material handling processes. A normal 3DP construction system can print a wall of a standard house of plinth area 120 m^2 in 2 days while the conventional process will complete the same in about 20 days, reducing the construction time by 90% and providing unmatched finishing (Zeltmann et al., 2016). The application of 3DP construction equipment may replace or reduce the finishing operations such as plastering, saving a lot of time, effort, and money. The resulting building is much sound, material distribution in homogenous and near defect-free. A large number of operations and processes are reduced or minimized, labor is almost eliminated, the probability of accidents and incidents due to human error and negligence are eliminated. The desired wall shapes may be conveniently developed including complex designs, contours, and profiles in three dimensions with variable wall thickness, etc. The use of auxiliary construction materials like shuttering is less and the waste is reduced to almost nil. The 3DP equipment is completely self-contained and autonomous, requiring 1–2 operators are the construction sites.

2.3   3DP Construction Process

A typical 3DP construction process is shown in Fig. 6. The construction may be divided into three stages as shown. The pre-processing includes needs analysis, selection of materials, creating a design using software, and its validation by stress analysis. The processing stage includes the transfer of the interchangeable design file in a compatible format to the 3DP, optimization of the model by simulation, and printing of the structure. The final stage is post-processing, which includes removing the shuttering or the support structure after curing, inspection, finishing, and repair/reworking if required. The entire process is automated and labor extensive.

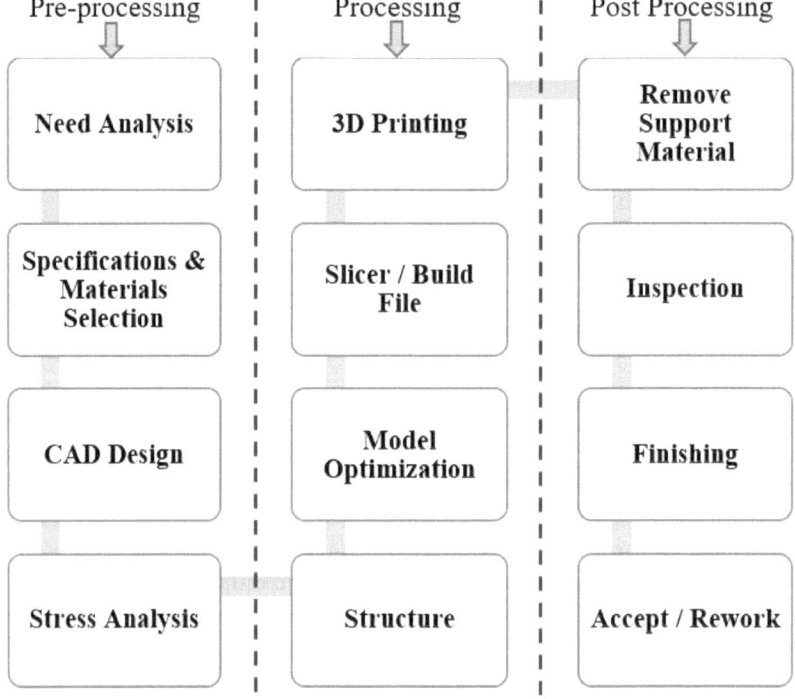

Fig. 6   A 3DP building construction process (*Source* Prepared by the authors)

3 Research Methodology

The research methodology of the present work includes the following. A comprehensive review of the applications of 3DP in construction across the globe and identifying the success stories. The estimation of the cost of construction and time required in printing the 3DP structures. The limitations and challenges in implementing the 3DP technology in construction. Analysis of the GDP of Oman and the contribution of infrastructure development and construction. The effect of 3DP construction on the cost, environment, time of construction, and estimation of the increment of the construction sector in the GDP of Oman. The cost–benefit analysis and cost estimation is done considering the regional, social, and demographic constraints in Oman.

4 Results

The 3DP construction technology includes building the walls, frames, and the slab but other parts such as the doors, windows, sanitary, and glass-work are to be fabricated separately. A 3DP technology to print the entire building with all its elements is yet to be achieved. Printing the huge structures, e.g., bridges, outdoor structures, etc., are typically very fast and completed with much ease. Printing of the prototypes is particularly very convenient, fast, and economical. The 3DP process includes pushing the premixed concrete or other building material through a nozzle in metered quantity and depositing it layer by layer. A few examples of the 3DP structures include contour crafting using concrete material. To increase the speed as well as to print the contours and complex profiles, more than one nozzle may be required. These nozzles may push the same or different construction materials with the same or variable speeds to achieve the designed profile. The conduits and accessories may also be printed simultaneously (The 13 Best Construction 3D Printers, 2019). A large-scale structure printing, a house built in 24 hours by the Russian company APIS (3D Printing: The Future of Construction, 2019), concrete printing, D-shape structure (Perkins & Skitmore, 2015), 3DP castle (Infra Construction and Equipment Magazine, 2019; Hager et al., 2016). The printing of concrete is an extrusion process where a premixed concrete slurry is continuously forced (extruded) through a nozzle and deposits layer by layer (Buswell et al., 2018; Paul et al., 2018; Perkins & Skitmore, 2015). Binder jetting and selective binding are achieved in 3DP

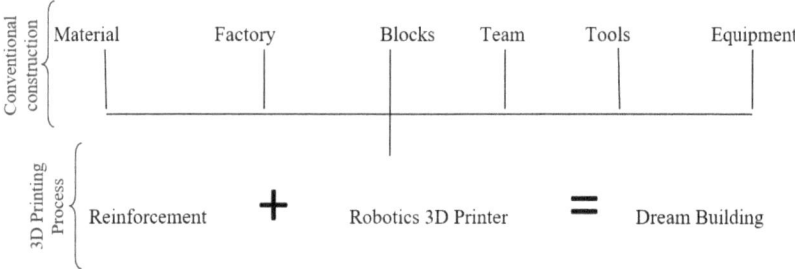

Fig. 7 A conventional construction process vs. a 3DP process (*Source* Prepared by the authors)

by a variable sprinkling of the binder over the solid particles at different locations. It uses the same principle as that in the inkjet printer (Ngo et al., 2018).

A comparison of a conventional construction process with a 3DP construction process is shown in Fig. 7. In the 3DP process, the stages of construction are reduced, wastage is minimized, and the process is fast.

The established 3DP materials used in building construction and their desirable properties are shown in Fig. 8 (Al-Qutaifi et al., 2018; Lu et al., 2019; Zhu et al., 2019). Other than concrete, a hybrid cellulose mixture is also being introduced as a promising biomaterial for construction (Mitchell et al., 2018).

The concrete should possess a very low slump and variable viscosity inside and outside the nozzle. The settling time of concrete should be quick enough to withstand its self-weight once it is layered. The ceramic powders namely, fly ash, plaster, mixed with polymer fibers and binders are becoming increasingly popular materials for 3DP. The use of such materials will not only reduce the weight of the structure and its cost but will also offer sustainable construction practices. Thermal insulation, green housing, affordable construction, and introduction of new construction materials hence may be promoted with 3DP (Nematollahi et al., 2018; Šerelis et al., 2016; Wangler & Flatt, 2019).

The use of 3DP in construction will open the pathways to explore new and smart construction materials with quick settling time. This technique has proven its worth for small-scale buildings and to print the construction elements, which may be assembled with the parent structure.

Fig. 8 3D printable
materials and desirable
properties (*Source*
Prepared by the authors)

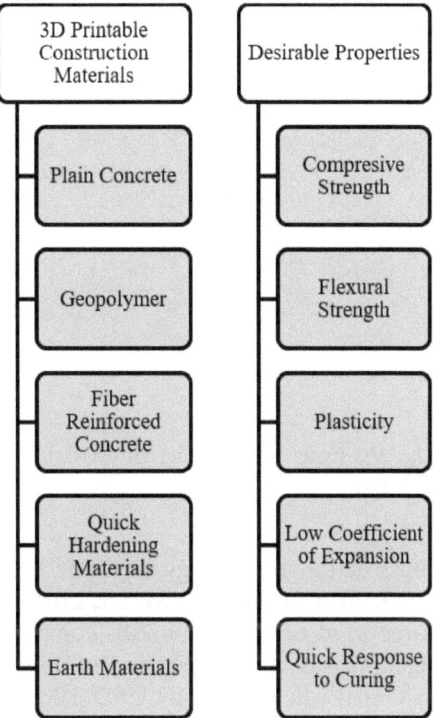

4.1   Cost–Benefit Analysis

Cost is the major issue in the application of the 3DP in construction but the centralized construction facilities and integration of electric, sanitary, and glasswork with the basic structure will make this process more economically feasible. In place of transporting the building material to the site of construction, if pre-fabricated blocks printed at a favorable site are transported and assembled at the construction site, the cost of 3DP can be made lower than the conventional construction methods (Thomas & Gilbert, 2015). Superior quality, time-saving, prevention of materials wastage, labor cost, and transportation costs are additional benefits. The cost and benefit analysis of a typical 3DP construction process is shown in Fig. 9.

For the cost–benefit analysis, three construction methods are being compared. The average cost of construction in Oman with conventional

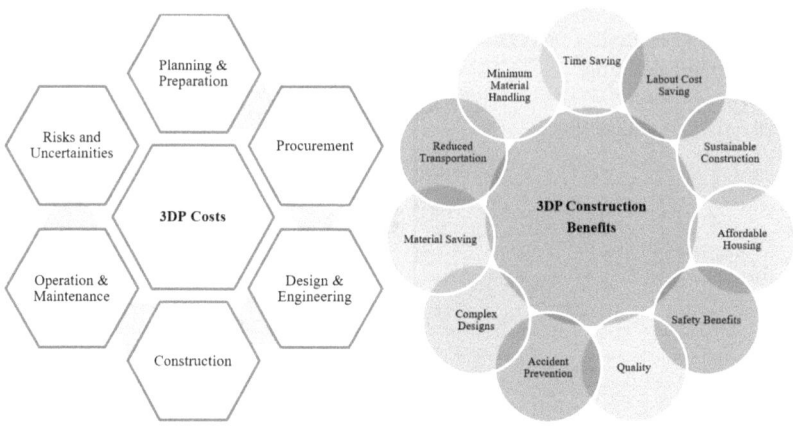

Fig. 9 Cost and benefits of 3DP in construction (*Source* Prepared by the authors)

construction technology is about USD 1337 per m² (Oman Projects, 2019; XE Currency Converter, 2019). The cost of modular construction is reported to reduce by 20% compared with the conventional methods (Hammad & Akbarnezhad, 2017). Therefore, the modular construction in Oman will cost about USD 1069.6 per m². As per the construction cost break-up published by APIS Company, Russia for the house shown in Fig. 7 (c) the price of the 3DP construction stages is given in Table 1 (All3DP, 2019).

Table 1 Price break-up of a 3D printed house in 2019

3DP construction stage (area 38 m²)	Price (USD)
Foundation	277
Walls	1624
Floor and roof	2434
Electric wiring	242
Doors and windows	3548
Finishing of the exterior	831
Finishing of the interior	1178
Total	10,134

Source Prepared by the authors based on data source from All3DP, 2019

Fig. 10 Customized 3DP construction technology for Oman (*Source* Prepared by the authors)

A proposed 3DP construction methodology for introduction in Oman is shown in Fig. 10. A central 3DP facility will print the modular parts of the building, such as slabs, blocks, fixtures, and support material. The modular parts will be printed as per pre-defined standard size based on the size of the building. The modular parts will be stored in the warehouse. When a small building, such as a house is required to be constructed, the modular components will be transported to the site of construction and assembled by the construction team. In case of multi-story buildings, such a method is expected to be commercially and economically feasible using earthmoving vehicles and construction equipment. The modular building parts and support material will be assembled floor by floor. In case a house or a building is required with a customized size or design, the design may be forwarded to the 3DP center and customized parts will be printed.

4.2   Construction Time

The time taken for construction of a house printed by the Russian company APIS (3D Printing: The Future of Construction, 2019) was 24 hours, which is at the rate of 1.58 m^2 per hour. The rate of construction for the same house by the modular method will be 0.0565 m^2 per hour and by a conventional method, it will be 0.0236 m^2 per hour (Hammad & Akbarnezhad, 2017; All3DP, 2019). Hence, it is inferred that the pace of construction by 3DP is 67 times faster than conventional construction and 28 times faster than modular construction.

The cost of construction by 3DP technology is USD 266.68 per m^2, which is only 20% of the cost of construction by conventional methods and 25% of the cost of construction by the modular construction method.

4.3   Impact on the Environment

The construction activities in the world are responsible for 11% of global carbon emissions (Finance and Commerce, 2019). The carbon emissions per m^2 from conventional construction is about 400 kg CO_2e (Klufallah et al., 2014). The introduction of modular construction is reported to save 15% of the material and the transportation cost. If the material saving alone is considered, there will be a reduction of 15% in the emissions (Lawson & Ogden, 2010), which is equivalent to 340 kg CO_2e per m^2 (Ranjha et al., 2018; Tirth et al., 2019). The emissions from 3DP construction methods are yet to be reported. Besides, in conventional construction, about 11% of the material is waste while in offsite (modular) construction, the material wastage is reduced to 1.8% (Waste Reduction, 2019). The material wastage in 3DP construction is considered negligible.

4.4   Impact on GDP

From the analysis of the different construction methodologies, it is learned that the 3DP construction is about 67 times faster than the conventional method and 28% faster than the modular construction. The cost of construction of 3DP construction per m^2 is only 20% of the conventional construction and 25% of the modular construction. The cost estimation is based on the cost quoted by a 3DP construction company, which includes the equipment and transportation cost (All3DP, 2019).

To estimate the impact of the introduction of 3DP in the construction sector in Oman, a simple consideration is made that 50% of the construction activities are completed by conventional methods and 50% by the offsite (modular) methods. The analysis of the data indicates that in Oman, the speed of infrastructure development will increase by about 50%, and within the same financial resource allocation, there may be an increase of 70% in the development of the real estate. Even after including several corrections, taxation, overheads, mobilization and logistics, and initial equipment costs, it is certain that 3DP has the potential to multiply the infrastructure development in Oman within the same financial resources. The contribution of infrastructure in the GDP may be doubled, which will improve the GDP as well as the HDI.

4.5   Limitations and Challenges in Using 3DP in Construction

Besides multiple advantages and cost benefits, the application of 3DP techniques in construction includes several limitations. The mold used for construction is difficult to separate and it often integrates and remains with the 3DP structure. The printing of concrete or other construction material has now been established and commercialized but the other components of a building such as plumbing, fixtures, services, wiring, flooring, woodwork, etc., are yet mature. The automated construction may not be efficient in large-scale production with a conventional design approach; hence, the design standards and construction practices require a revamp (Yossef & Chen, 2015). Presently, only a few construction materials are compatible with the 3DP machines, a new material may require redesign and fabrication of the 3DP machine. The cost of the initial investment in procuring the 3DP machines may not be economically feasible for small-scale construction companies.

5   CONCLUDING REMARKS

New technologies are being introduced continuously with time. Every new technology has a golden period for adoption, before which, it is expensive and after which the benefits saturate and it becomes obsolete. Infrastructure development is a key contributor to the GDP and the HDI of the country. It is a direct contributor to the GDP of a country and affects the other components indirectly. The fast-developing economies of the world have more than 5% contribution to infrastructure development (construction) to the GDP while Oman's status has been less than 2%. To improve economic growth, the contribution of infrastructure to the total GDP has to be increased.

The introduction of 3DP in the construction sector in Oman has the potential of increasing the pace of infrastructure development by conventional methods by about 67 times and modular construction by about 28 times. The cost of 3DP structures is much lesser and the material wastage is almost nil. In the same budget, the 3DP construction may increase the real estate output by 75% and at a faster pace. The impact of

3DP construction on emissions and the environment is highly favorable. For a progressive developing nation like Oman, the introduction of 3DP construction technology is inevitable. The time for the introduction of the 3DP construction has come in Oman and it should be done without any delay.

REFERENCES

3D Printing: The Future of Construction. (2019). https://www.3dnatives.com/en/3d-printing-construction-310120184/.

Aimar, A., Palermo, A., & Innocenti, B. (2019). The role of 3D printing in medical applications: A state of the art. *Journal of Healthcare Engineering, 2019.* https://doi.org/10.1155/2019/5340616.

All3DP how much does a 3D printed house coast in 2019? (2019). https://all3dp.com/2/3d-printed-house-cost/.

Al-Qutaifi, S., Nazari, A., & Bagheri, A. (2018). Mechanical properties of layered geopolymer structures applicable in concrete 3D-printing. *Construction and Building Materials, 176,* 690–699. https://doi.org/10.1016/j.conbuildmat.2018.04.195.

Business Live ME. (2019). https://www.businessliveme.com/economy/infrastructure/industry-insights-omans-infrastructure-development-2019/.

Buswell, R. A., Leal de Silv, W. R., Jones, S. Z., & Dirrenberger, J. (2018, June). 3D printing using concrete extrusion: A roadmap for research. *Cement and Concrete Research, 112,* 37–49. https://doi.org/10.1016/j.cemconres.2018.05.006.

Finance and Commerce, 2019. (2019). https://finance-commerce.com/2019/06/the-carbon-footprint-of-modern-construction-is-huge/.

García de Soto, B., Agustí-Juan, I., Hunhevicz, J., Joss, S., Graser, K., Habert, G., & Adey, B. T. (2018, May). Productivity of digital fabrication in construction: Cost and time analysis of a robotically built wall. *Automation in Construction, 92,* 297–311. https://doi.org/10.1016/j.autcon.2018.04.004.

Hager, I., Golonka, A., & Putanowicz, R. (2016). 3D printing of buildings and building components as the future of sustainable construction? *Procedia Engineering, 151,* 292–299. https://doi.org/10.1016/j.proeng.2016.07.357.

Hammad, A. W. A., & Akbarnezhad, A. (2017). Modular vs. conventional construction: A multi-criteria framework approach. In *ISARC 2017—Proceedings of the 34th International Symposium on Automation and Robotics in Construction* (Isarc) (pp. 214–220). https://doi.org/10.22260/isarc2017/0029.

Infra Construction and Equipment Magazine. (2019). https://www.nbmcw. com/report/construction-infra-industry/39828-3-d-printing-in-construct ion-a-review.html.

Kidwell, J. (2017). *Best practices and applications of 3D printing in the construction industry* (pp. 1–8). https://digitalcommons.calpoly.edu/cgi/vie wcontent.cgi?article=1090&context=cmsp.

Klufallah, M. M. A., Nuruddin, M. F., Khamidi, M. F., & Jamaludin, N. (2014). Assessment of carbon emission reduction for buildings projects in Malaysia— A comparative analysis. In *E3S Web of Conferences, 3*. https://doi.org/10. 1051/e3sconf/20140301016.

Lawson, R. M., & Ogden, R. G. (2010). Sustainability and process benefits of modular construction. In *8th CIB World Building Congress* (pp. 38–51). https://www.irbnet.de/daten/iconda/CIB18783.pdf..

Lu, B., Weng, Y., Li, M., Qian, Y., Leong, K. F., Tan, M. J., & Qian, S. (2019). A systematic review of 3D printable cementitious materials. *Construction and Building Materials, 207*, 477–490. https://doi.org/10.1016/j.conbuildmat. 2019.02.144.

Mitchell, A., Lafont, U., Hołyńska, M., & Semprimoschnig, C. (2018). Additive manufacturing—A review of 4D printing and future applications. *Additive Manufacturing, 24*(September 2017), 606–626. https://doi.org/10.1016/ j.addma.2018.10.038.

Nematollahi, B., Vijay, P., Sanjayan, J., Nazari, A., Xia, M., Nerella, V. N., & Mechtcherine, V. (2018). Effect of polypropylene fibre addition on properties of geopolymers made by 3D printing for digital construction. *Materials, 11*(12). https://doi.org/10.3390/ma11122352.

Ngo, T. D., Kashani, A., Imbalzano, G., Nguyen, K. T. Q., & Hui, D. (2018). Additive manufacturing (3D printing): A review of materials, methods, applications and challenges. *Composites Part B: Engineering, 143*(December 2017), 172–196. https://doi.org/10.1016/j.compositesb.2018.02.012.

Oman Projects, 2019. (2019). https://www.google.com/search?q=cost+of+con struction+in+oman&rlz=1C1GGRV_enSA752SA753&oq=cost+of+construct ion+in+oman&aqs=chrome..69i57.6951j1j8&sourceid=chrome&ie=UTF-8.

Oman Vision 2040. (2019). https://www.2040.om/en/#2040Themes.

Paoletti, I., & Ceccon, L. (2018). The evolution of 3D printing in AEC: From experimental to consolidated techniques. *3D Printing*. https://doi.org/10. 5772/intechopen.79668.

Paul, S. C., van Zijl, G. P. A. G., & Gibson, I. (2018). A review of 3D concrete printing systems and materials properties: Current status and future research prospects. *Rapid Prototyping Journal, 24*(4), 784–798. https://doi.org/10. 1108/RPJ-09-2016-0154.

Perkins, I., & Skitmore, M. (2015). Three-dimensional printing in the construction industry: A review. *International Journal of Construction Management, 15*(1), 1–9. https://doi.org/10.1080/15623599.2015.1012136.

Ranjha, S., Kulkarni, A., & Sanjayan, J. (2018, November). 3D construction printing—A review with contemporary method of decarbonisation and cost benefit analysis. In *1st international conference on 3D construction printing (3DcP)* (pp. 1–11).

Šerelis, E., Vaitkevičius, V., & Keršzevičius, V. (2016). Mechanical properties and microstructural investigation of ultra-high performance glass powder concrete. *Journal of Sustainable Architecture and Civil Engineering, 14*(1), 5–11. https://doi.org/10.5755/j01.sace.14.1.14478.

Statista. (2019). https://www.statista.com/statistics/566787/average-yearly-expenditure-on-economic-infrastructure-as-percent-of-gdp-worldwide-by-country/.

The 13 best construction 3D printers in 2019. (2019). https://www.aniwaa.com/house-3d-printer-construction/.

Thomas, D. S., & Gilbert, S. W. (2015). Costs and cost effectiveness of additive manufacturing: A literature review and discussion. In *Additive manufacturing: Costs, cost effectiveness and industry economics* (pp. 1–96). Nova Science Publishers.

Tirth, V., Algarni, S., Agarwal, N., & Saxena, A. (2019). Greenhouse gas emissions due to the construction of residential buildings in Moradabad, India. *Applied Ecology and Environmental Research, 17*(5), 12111–12126. https://doi.org/10.15666/aeer/1705_1211112126.

Trading Economics. (2019). https://tradingeconomics.com/oman/gdp-growth-annual.

UNDP. (2018). Human development indices and indicators: 2018 statistical update. *United Nations Development Programme, 27*(4), 123. Retrieved from http://hdr.undp.org/en/content/human-development-indices-indicators-2018-statisticalupdate#text=Human%20Development%20Indices%20and%20Indicators%3A%202018%20Statistical%20update%20is%20being,trends%20in%20human%20development%20indicators.

Wangler, T., & Flatt, R. J. (2019). *Correction to: First RILEM International conference on concrete and digital fabrication—Digital concrete 2018*. https://doi.org/10.1007/978-3-319-99519-9_31.

Waste Reduction Potential of Offsite Volumetric Construction. (2019). http://www.wrap.org.uk/sites/files/wrap/VOLUMETRIC%20-%20Full%20case%20study.pdf.

Worldometers Oman Population. (2019). https://www.worldometers.info/world-population/oman-population/.

XE Currency Converter. (2019). https://www.xe.com/currencyconverter/convert/?Amount=1&From=OMR&To=USD.

Yossef, M., & Chen, A. (2015). Applicability and limitations of 3D printing for civil structures. In *Proceedings of the 2015 Conference on Autonomous and Robotic Construction of Infrastructure* (pp. 1–10). Ames, Iowa: Iowa State University. https://www.researchgate.net/publication/277665549_Applicabi lity_and_Limitations_of_3D_Printing_for_Civil_Structures.

Zeltmann, S. E., Gupta, N., Tsoutsos, N. G., Maniatakos, M., Rajendran, J., & Karri, R. (2016). Manufacturing and security challenges in 3D printing. *JOM Journal of the Minerals Metals and Materials Society, 68*(7), 1872–1881. https://doi.org/10.1007/s11837-016-1937-7.

Zhu, B., Pan, J., Nematollahi, B., Zhou, Z., Zhang, Y., & Sanjayan, J. (2019). Development of 3D printable engineered cementitious composites with ultra-high tensile ductility for digital construction. *Materials and Design, 181,* 108088. https://doi.org/10.1016/j.matdes.2019.108088.

Driving Factors of Adopting 4.0 IR Technologies in the Logistics Sector

Zainab Al Balushi, Anwar Al Sheyadi, and Ali Al Shidhani

1 INTRODUCTION

In today's economy, all business activities and practices are constantly adjusted in response to social, environmental, and technology requirements. Those changes drive new business models and innovative modes of operating supply chains that are empowered by devices and technologies which can autonomously interact with each other and amplify value creation among members throughout the value chain. Such changes should be addressed by decision-makers through developing the required

Z. Al Balushi (✉)
College of Economics and Political Science, Sultan Qaboos
University, Muscat, Sultanate of Oman
e-mail: zeinaba@squ.edu.om

A. Al Sheyadi
College of Applied Sciences, University of Technology and Applied Sciences,
Rustaq, Sultanate of Oman
e-mail: anwar.alsheyadi.rus@cas.edu.om

A. Al Shidhani
Ministry of Transportation and Communication and Information Technology,
Muscat, Sultanate of Oman
e-mail: ali.alshidhani@mtcit.gov.om

© The Author(s), under exclusive license to Springer Nature   109
Singapore Pte Ltd. 2021
N. R. Al Mawali et al. (eds.), *Fourth Industrial Revolution and Business
Dynamics*, https://doi.org/10.1007/978-981-16-3250-1_6

information systems, technology, physical facilities that are capable to meet emerging social, economic, and environmental needs. Accordingly, in the logistics context, there is a need for a new paradigm of developing logistics systems and how globally freights are supplied, moved, stored, and realized. 4.0 IR technologies and applications can be seen as effective tools to boost global logistics systems with the competencies required to manage contemporary business models and integrated supply chains. Hence, logistics represents an appropriate application area for 4.0 IR. The new technological advancement and the growing complexity in logistics operations are considered as a critical topic due to the resulted implications of these advancements and consequent convolution. The concept of Logistics 4.0, which describes the adoption of 4.0 IR in the logistics industry, is still evolving and it lacks a clear understanding. It is not yet fully established in practice, and its future impacts on logistics management are rather uncertain (Hofmann & Rüsch, 2017). The literature has recognized the major impacts of the digital transformation of transportations and logistics activities on cost reduction, customer satisfaction as well as on environmental and social improvements (Kuhlmann & Klumpp, 2017). Although evolving technological changes have and will greatly affect firms' decision to invest in the development and adoption of 4.0 IR, there is hardly rare attention is being given to understanding the actual main drivers behind logistics firms' decision to invest and in what technology. Although technological, economic, and environmental consequences of the 4.0 IR in logistics sectors are significantly addressed in the literature, the drivers of being part of this revolution are not adequately addressed, especially in the context of less developed countries such as Oman. Understanding these driving forces is important especially in such context. This could potentially guide firms to understand the factors which are positively influencing the adoption of 4.0 IR technologies and the lack of perceived benefits associated with Logistics 4.0.

This study aims at providing an overview of the current situation of 4.0 IR in Omani logistics industry, to explore the main factors that drive the decision to adopt these technologies and to investigate the mechanisms through which drivers and practices of 4.0 IR interacts. To achieve the research objectives, the paper is structured as following. In the next section, an overview of the literature on drivers and trends of using 4.0 IR in logistics sectors and the subsequent development of the research hypotheses are provided. Then, a discussion on the methods used for data collection and data analysis is included in Sect. 3, which is followed by the results of data analysis. Next, discussion and research implications are discussed. Finally, limitation and general discussion on future research areas are highlighted.

2   Literature Review

2.1   Logistics 4.0 in the Local Context

The Sultanate of Oman has set a target to diversify its economy further than dependency on hydrocarbon resources. Thus, it identified several industries of great potential such as aviation, fisheries, mining, and logistics. Oman enjoys unique and strategic geopolitical alliances which makes it a perfect destination and en route for international trade. As of 2019, the logistics sector contributes 3.75% of the country's GDP. The ambition is to rapidly increase the sector's contribution in the coming few years. Technology and innovation are disrupting every industry; logistics is no different. Smart sensors collecting data transferred over high-speed networks stored in the cloud enables an array of new applications and use cases in the logistics sector. Based on Hofmann and Rüsch (2017) logistics-oriented 4.0 IR application model, Logistics 4.0 has a significant role in optimizing value creation in intefirm logistics, though enabling real-time information flows, allowing end-to-end supply chain transparency, and improving flexibility.

Emerging technologies such as Artificial Intelligence (AI) and Distributed Ledger Technologies (DLTs) have the potential to revolutionize the logistics industry by offering more operational efficiencies, lower latency, and new revenue streams. AI is currently transforming the online shopping experience where customers are guided toward making specific purchases based on customer's preferences and behavior. Last mile delivery solutions are also making use of AI to predict future demand and aggregating shipments and provide route efficiencies. Furthermore, deliveries in difficult terrains are handled by Unmanned Arial Vehicles (UAV). DLTs have a potential usage in supply chain management to trace goods from source to destination. A combination of AI, DLT, and UAV can shape the future of last mile delivery. The abovementioned technologies, in addition to others such as Internet of Things (IoT), 3D Printing, and Virtual Reality (VR), are collectively known as the Logistics 4.0 technologies. It is inevitable for Logistics Service Providers to incorporate the use of 4.0 IR technologies in the business strategy to edge over competitors, allowing optimization of value chains, mass transportation, and real-time coordination.

2.2   Trends of 4.0 IR in Logistics and Transportation Industries: An Overview

The previous three industrial revolutions permanently changed global economic systems, and they lead to the creation of the 4.0 IR. Several definitions and terms have been used for the 4.0 IR, all of which focus on concepts and technologies of value chain organization that simultaneously uses the Cyber-physical systems (CPS), Big Data, Internet of Things (IoT), Internet of Services (IoS), and smart factory (Hofmann & Rüsch, 2017). It refers to a further development stage in the processes of managing and organizing the entire value chain activities involved in the production of products or services (Maslarić et al., 2016). It is made up of interconnected systems of machines, sensors, software, workpieces, and communication technologies that are used to create virtual copy of the physical world, monitor physical operations, enable more decentralized decisions to be made faster, and allow real-time response to systems and market changes (Khaitan & McCalley, 2015). These technologies communicate with each other and with other physical and virtual production resources in real time as products and services are been produced for customers. With these technological changes, improvement in design, sourcing, production, storage, distribution, and exploitation areas are expected to be achieved, which in turn will have clear impacts on all areas of logistics including the transformation of all goods, financial, and information flows.

As a function, logistics exists in all phases of industry development. Yet, its purpose and modes of realizing its activities have changed over the past few years to reflect the constant increase in its complexity which mainly resulted from the ongoing increasing demand for greater efficiency and highly customized customer services. So, there is a need for logistics organizations to look for more efficient solutions that meet the current and future challenges in industrial practice in which 4.0 IR, 4.0 can play a critical role to achieve these objectives. There are several examples of Logistics 4.0 machines, tools, and applications that can be used to achieve and conduct self-organized logistics processes that meet expectations of all supply chain members. When considering the current trends of using 4.0

IR in logistics processes, these technologies can be generally classified into three categories: networking, IT and software development, and robotic and sensor technology (Parham & Tamminga, 2018).

Logistics ought to be a network where all processes are connected with each other in an interactive manner to enhance analytical potentialities. With the viability of data and the utilization of IT, many firms are able to optimize the use of existing resources. Present "approaches" of value creation struggle to cope with the increasing pressures to attain sustainability, cost efficiency, adaptability, flexibility, and stability (Hofmann & Rüsch, 2017) that are critical in managing global supply networks. These potentials can be captured through the adoption of 4.0 IR technologies. Winkelhaus and Grosse (2019, p. 4) define Logistics 4.0 as "the logistical system that enables the sustainable satisfaction of individualized customer demands without an increase in costs and supports this development in industry and trade using digital technologies". Yet the prerequisites for utilizing 4.0 IR in the logistics sector are digitalization of processes that leads to "logistics transformation" or to creation of "smart logistics" solutions (Maslarić et al., 2016) besides workforce capabilities and education (Hofmann & Rüsch, 2017). Furthermore, according to Barreto et al. (2017), Logistics 4.0 paradigm is mainly optimizing inbound and outbound logistics processes robotically through fully automated and intelligent systems, embedded in software and databases from which required data is stored, analyzed and communicated through Internet of Things (IoT) systems. Hence, it is expected to enable analyzing, communicating, designing, understanding, and optimizing real-time tracking of resource flows, value-added transport handling as well as an precise risk management (Hofmann & Rüsch, 2017). Moreover, with the increased complexity of managing logistics, expected potentials should be evaluated situationally as logistics firms' decision to invest in such technologies are driven by internal and external factors. Hence, this study investigates the factors that influence logistics firms' decision to adopt 4.0 IR technologies from institutional theory lens, which is expected to contribute to the emerging literature of Logistics 4.0 as well as the literature of the institutional theory.

2.3   Drivers of Adopting 4.0 IR in Logistics Sectors: An Institutional Theory Perspective

This study aims to investigate the main drivers that encourage or sometimes force the growing adoption of 4.0 IR technologies, tools, and applications among logistics firms. The institutional theory is used as the main theoretical instrument to explore these drivers and explain the process through which they can influence on firms' decisions to adopt 4.0 IR technologies. The institutional theory argues that organizations adapt their business practices as doing so enhances their legitimacy (DiMaggio & Powell, 1983). Organizational structures, operations, behaviors, and practices are steered by social expectations and norms of the institutional environment (DiMaggio & Powell, 1983; Scott, 2008; Zucker, 1987) which collectively generate enough institutional pressures to drive organizational decisions in different directions. The institutional forces can generally be generated internally within the firm or externally in the exchange environment to encourage convergent business practices (Zsidisin et al., 2005). According to DiMaggio and Powell (1983), the institutional forces can be classified into normative, mimetic, and coercive pressures, which are discussed below.

Normative pressures are the outcomes of mutual expectations within particular organizational contexts of what composes appropriate behavior (DiMaggio & Powell, 1983). These pressures force organizations to adopt techniques that are considered effective by the community of practice, such as professional standards and practices established by education and training methods, professional networks, and the movement of employees among firms (Sherer et al., 2016). On the other side, mimetic and coercive pressures are driven mainly by external parties. *Mimetic pressures* originate from a firm's perceived success of competitors' actions. They emerge from uncertainty that encourages imitation (Zsidisin et al., 2005) and, thus, they force organizations to replicate other successful organizations when there is limited conception of a new process, technology, or external conditions (Sherer et al., 2016). Benchmarking is a typical approach used by firms to reflect the mimetic pressures they face (Liu et al., 2010). In contrast, *coercive pressures* instigated from political influences exerted by the powerful exchange partner in the supply chain (i.e., supplier or a customer) on which the focal firm depends. Coercive pressures are derived by resource-dominant organizations, regulatory bodies, market forces such as competition, and/or entities that have resources on which an organization depends (Sherer et al., 2016). With the coercive pressures firms are forced to serve interests of other powerful firms by demanding the implementation of their favorable operational practices (Liu et al., 2010).

The institutional theory perspective has been used by previous Logistics and Supply Chain Management (SCM) studies to explain the main reasons for organizational decisions, actions, and activities. Kauppi (2013) argues that three sub-areas are adopting institutional theory more than others, mainly, (1) technology applications in logistics and supply chain management; (2) explaining environmental-related practices in SCM, and (3) quality programs. This theory has also been used to examine factors that motivate logistics service providers to develop green capabilities and enhance environmental performance (e.g., Chu et al., 2018, 2019; Raitasuo et al., 2019). Also, it has been used to study ports logistics integration (e.g., Alavi et al., 2018), logistics outsourcing (e.g., Wan et al., 2019), and humanitarian logistics (e.g., Oloruntoba, 2018). However, none of the existing studies have empirically investigated how and the extent to which various institutional forces affect firms' decision to adopt new innovation in logistics sectors such as the adoption of Logistics 4.0 technologies, tools, and applications. From a theory building perspective, Barratt and Choi (2007) use case studies to investigate decentralized business units' responses to pressures to adopt radio frequency identification (RFID). Seven propositions evolved from the study addressing institutional rationalization, technical rationalization, perceived uncertainty, isomorphism, and decoupling. Liu et al. (2010) investigate how institutional pressures motivate the firm to adopt Internet-enabled Supply Chain Management systems (eSCM) and how such effects are moderated by organizational culture. They found that the different institutional pressures have differential direct effects on firms' eSCM adoption intention: normative and coercive pressures are positively associated with adoption intention, mimetic is not. Zsidisin et al. (2005) integrated both open systems theory and institutional theory to analyze the creation of business continuity plans for purchasing, and develop propositions on how isomorphic pressures lead firms to similar risk management practices over time. They found that the three pressures of regulating, validating, and habitualizing all influence how firms ensure continuity in their supply chains.

Prior to investing in and adopting 4.0 IR innovative solutions (like Logistics 4.0 in our case), organizations seek to obtain information about institutional norms and expectations (Liu et al., 2010). Using these information, organization will be able to conduct cost–benefit analysis, and decide on hedging against pontifical uncertainties (Scott, 2008). In addition to this coercive pressure, in industries dominated with few large customers adopting certain technology, their supply chain parties perceive adoption of this technology as mandatory to maintain their relation even if they were not forced to do so (Narayanan et al., 2009). Thus, organization decision to invest in interorganizational information systems are driven by normative pressure (Teo et al., 2003).

Gupta and George (2016) argue that, top managers with management and IT education background impose data-driven decision-making. Based on their empirical study, Dubey et al. (2019) found that organization selection of technological innovations resources is significantly influenced by institutional pressures. The above discussion reveals that in general normative, coercive, and mimetic pressure would influence logistics firms to adopt various technologies such as those related to Logistics 4.0, and thus the following hypotheses are proposed:

H 1: A firm's perceived normative pressures toward Logistics 4.0 adoption are positively related to its Logistics 4.0 adoption.

H 2: A firm's perceived coercive pressures toward Logistics 4.0 adoption are positively related to its Logistics 4.0 adoption.

H 3: A firm's perceived mimetic pressures toward Logistics 4.0 adoption are positively related to its Logistics 4.0 adoption.

Finally, we used firm size and years in business as control variables in our model because the literature suggested that firm characteristics such as size and age could influence the level of innovation used by firms (Liu et al., 2010).

3   RESEARCH METHODOLOGY

3.1   Data Collection

The data were collected through a survey administrated to a random sample of Omani logistics firms. The questionnaire included 16 items related to the extent to which the firm adopted 4.0 IR technologies, tools, and application to conduct their various logistics processes and the extent to which adopting the practices have been influenced by the pressure of various institutional forces.

Item used to develop the instrument were measured using a 5 point Likert scale. These items were mainly adopted from SCM and logistics literatures (e.g., Khalifa & Davison, 2006; Teo et al., 2003) and based on experts' suggestions. Normative pressures were measured by three items on the extent of 4.0 IR adoption by business partners, customers, and shareholders of the focal firm; mimetic pressures were measured by two items on the perceived success of competitors and industry leaders that had adopted Logistics 4.0; coercive pressures were measured by three items. 4.0 IR technologies, tools, and applications adoption among the target companies were measured using eight items on the level of adoption of these technologies. The first draft of the questionnaire was modified based on comments and suggestions received from academic and industry logistics experts on its validity, clarity, and suitability to the context of the study, and it was then piloted on 4 Omani logistics firms.

Conducted during the period of August 2019 till January 2020, the survey was mailed to a total of 350 logistics firms. As follow-up, a reminder letter was sent to non-respondents. We received a total of 104 completed questionnaires. This yielded a total response rate of approximately 29.7%. For the non-responding firms, and after calling some of these firms, the major reason for not responding was found to be the firm's policy not to participate in any academic survey invitations.

The survey respondents had senior to middle management experience and majority of the respondents had more than 8 years work experience in their firm, suggesting they are appropriate respondents to provide the information needed. Considering the characteristics of the participated firms, results of the descriptive statistics show that the median number of full-time employees in the responding companies was 93 and more than 68% of the firms have more than 8 years in business.

3.2 Validity and Reliability Tests Results

Given that all the data used in this study were collected from a single respondent within each single firm, and some of the targeted firms decided not to respond to the questionnaire, non-response bias, and common method bias may present problems in our data analysis (Hair et al., 2010; Liu et al., 2010). Using suggestions of previous studies, we evaluated the non-response bias by performing an independent t-test to determine if there were significant differences in the mean values of responses on size (number of full-time employees) of the early set of respondents and the late set of respondents (Armstrong & Overton, 1977). The t-test revealed no significant differences ($p < 0.05$) between the two groups of the respondents on size. Furthermore, Harman's single-factor test in which unrotated factor analysis with eigenvalue greater than 1 was used to evaluate the common method bias. According to this test, and after entering all of the measurement items into a single-factor test, common method bias exists if the items load on a single factors or if any specific factor explains most of the variance (Liu et al., 2010). Results of this test showed the presence of four different factors and that the first factor explains only 27% of the variance in the data, revealing that common method bias is not a critical issue in our data (Hair et al., 2010).

Concerning the reliability and validity of our constructs, as illustrated in Table 1, reliability analysis revealed the Composite Reliability for this questionnaire ranging from 0.70 to 0.92 for different constructs, which considered to be internally consistent (Hair et al., 2010). Further, validity analysis showed that factorial weights for all indicators were greater than 0.5 ($p < 0.05$), ranging from 0.645 to 0.860, and the Average Variance Extracted (AVE) for all constructs was above 0.5, revealing a good level of convergent validity (Fornell & Larcker, 1981). The discriminant validity of the constructs was assessed by comparing the square root of AVE of the constructs with the correlations between any pair of the constructs. As illustrated in Table 2, the square root of AVE for our constructs is greater than the correlation between the constructs and the other constructs, suggesting a good evidence of discriminate validity (Fornell & Larcker, 1981).

3.3   Results of Data Analysis

Confirmatory factors analysis by the mean of Structural Equation Modeling (SEM) was used to empirically test our conceptual model and our hypotheses. All of the descriptive analyses were conducted using SPSS 21.0 and the inferential analyses were conducted using AMOS 21.0 with a maximum likelihood estimation. In order to test the possible effect of the institutional pressure on firms' willingness to adopt various 4.0 IR technologies, tools, and applications for conducting various logistics activities, we created a model that includes the direct links of the three forms of institutional pressure to different types of 4.0 IR technologies, tools, and applications (Fig. 1). We assessed the model in terms of the overall

Table 1   Descriptive statistics for constructs used in the current study

Construct (Label)	Measurement items	Mean/S.D	Factorial weight	CR	AVE
Normative pressure (Norm)	Customers	4.30/1.24	0.849	0.84	0.64
	Shareholders	3.12/0.98	0.693		
	Business partners	3.48/1.06	0.809		
	Average	3.63/1.09			
Corecive pressure (Core)	Non-government organizations	3.03/0.91	0.707	0.81	0.60
		3.36/1.02	. 823		
	Government	3.29/0.94	0.781		
	Society	3.23/0.96			
	Average				
Memetic pressure (Mem)	Competitors pressure	4.48/1.11	0.704	0.70	0.54
	Industry pressure	3.97/1.21	0.766		
	Average	4.23/1.16			
Logistics 4.0 technologies, tools and applications Log.4)	Cyber_security_solutions	4.14/2.03	0.832	0.92	0.58
	Internet_of_Things	4.24/2.12	0.860		
	Cloud logistics	3.52/2.02	0.815		
	Big data and artificial intelligence	3.87/2.21	0.645		
	RFID technologies	3.72/2.00	0.803		
	ERP (Enterprise resource planning)	4.15/1.78	0.740		
	IVMS or track and trace systems	3.93/2.03	0.649		
	Supervisory control and data acquisition (SCADA) systems	*3.71/1.06*	0.701		
	Average	3.91/1.91			

Source Survey data conducted during August 2019 till January 2020 in Oman

Table 2   Discriminant validity test

	Norm	Core	Mem	Log.4
Norm	**0.80**			
Core	0.217	**0.78**		
Mem	0.301	0.284	**0.73**	
Log.4	0.374	0.153	0.461	**0.76**

Source Survey data. *Note* Square root of AVE in bold on diagonal, Correlation is significant at $p < 0.05$ for values greater than 0.160, at $p < 0.01$ for values greater than 0.200, at $p < 0.001$ for values greater than 0.315, $N = 104$

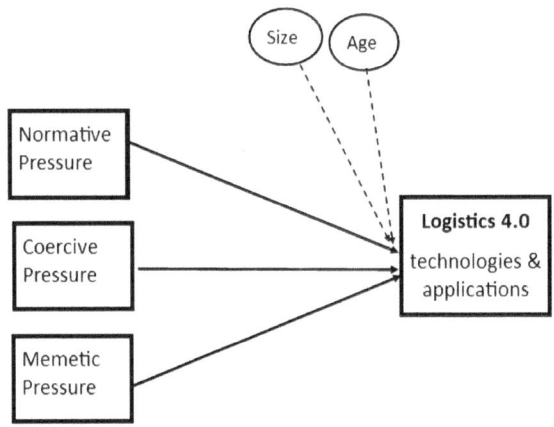

Fig. 1   Conceptual model of the study (*Source* Developed by researchers)

model fit and assessed the size, direction, and significance of structural path coefficients of the model (Hair et al., 2010).

Concerning the model fit assessment, results (Table 3) reveal that our model has achieved an acceptable overall fit to the data. Further, concerning hypotheses testing, results of the SEM (Table 4) showed that the relationships between Logistics 4.0 and normative pressure (β = 0.378, p < 0.001), and memetic pressure (β = 0.675, p < 0.001) were strongly significant, which strongly support H1 and H3, respectively. However, these results revealed that the relationship between Logistics 4.0 and coercive pressure (β = 0. 125) was not significant, and thus H2 is not supported. The control variable of firm size was found to have positive and significant effect on Logistics 4.0 adoption (b = 0. 174, p

Table 3   Structural models good of fit results

Models\indices	χ^2 (df)	Normed χ^2	GFI	CFI	IFI	RMSEA	PNFI
Model 1	369.8(162)	2.283	0.825	0.935	0.938	0.078	0.784
Recommended values (Hair et al., 2010)	NA	< 3.0	≥ 0.8	≥ 0.9	≥ 0.9	< 0.10	≥ 0.70

Source Survey data

Table 4   Results of structural equation modeling

Structural paths	Standardized estimates	t-value	Results
Normative→Log.4	0.378***	5.583	H1 supported
Coercive→Log.4	0.125	1.426	H2 not supported
Memetic→Log.4	0.675***	7.120	H3 supported
Control variables	0.174*		
Size→Log.4	0.088		
Age→Log.4			

Source Survey data, *Path is significant at $p < 0.05$ for values greater than 0.170, $p < 0.01$ for values greater than 0.250, at $p < 0.001$ for values greater than 0.300

< 0.05), but firm age does not have a significant effect on Logistics 4.0 adoption ($b = 0.088$, $p = 0.155$). The implications of our findings will be discussed in the following section.

4   Results

Customers' requirements, competitors' actions, and government expectations institute normative, memetic, and coercive pressures. The empirical results of this study reveal that not all of these sources of institutional pressures have a positive significant effect on logistics firms' adoption of 4.0 IR technologies. According to the arguments of the institutional theory and findings of previous researches on logistics innovation, this study posits that due to industry, customers, society expectations normative, memetic, and coercive pressures should have positive direct effects on firms' development and deployment of 4.0 IR technologies. These relationships were expected to vary depending on firms' characteristics including size and age of the firm. Despite the growing recognition about the importance of 4.0 IR in achieving higher performance, arguments of previous studies on main factors that drive the growing adoption of 4.0 IR are inconclusive. Accordingly, for the sake of gaining a greater understanding of the current status of adopting 4.0 IR among Omani logistics firms and to investigate the role of institutional forces in driving this adoption, this study was conducted.

Findings of the descriptive statistics of this study reveal that, in general, logistics firms in Oman are still at the early stages of adopting many of 4.0 IR technologies, applications, and tools. Emerging technologies such as IoT, cloud logistics, cybersecurity solutions, big data, and AI are getting widely recognized by Omani logistics firms, which can be seen from the mean value of firms' responses on the extent to which they adopt these technologies.

From the structural analysis results, findings of this study support most of these assumptions, and thus showing that institutional pressure would encourage higher level of 4.0 IR adoption among Omani logistics firms. Our structural analysis results show that the emergence of these technologies with the growing mimetic pressure were the most critical forces pushing logistics firms to expand their commitment toward adoption of 4.0 IR. For the sake of achieving higher levels of efficiency and gaining higher market shares, many logistics firms have already implemented 4.0 IR technologies. This is expected since the competition level in the Omani logistics sectors is increasing, especially from the European and South Asian logistics provided which are already further ahead in implementing 4.0 IR than many logistics provided firms from other countries. Normative pressures were also found to be critical factor in 4.0 IR adoption among Omani logistics firms, which may show the gradual shift in the mind of the consumer and other business partners, demanding for more flexible and highly responsive production of logistics services. This demand from customers is driven by the proliferation of mobile internet with just-in-time services in various other sectors such as entertainment which improved customer experience. Therefore, customers expect logistics service providers to deliver similar experience levels.

However, the assumption regarding the role of coercive pressure on 4.0 IR adoption was not supported. The possible reasons for the later finding may be because the international nature of the logistics industries and the international focus of Omani logistics firms. This may show that local regulatory and social pressures are not considered yet as critical forces to push further adoption of 4.0 IR among Omani logistics firms, and thus more market, rather than regulatory, incentives are needed to encourage further adoption of 4.0 IR. From the institutional theory perspective, firms normally tend to prioritize stakeholders' requirements depending on the power of stakeholders, where demands of the most important stakeholder such as the competitors and customers will be given the highest priority and lesser or no attention will be given to the demands of less important stakeholders (DiMaggio & Powell, 1983; Kauppi, 2013; Zucker, 1987).

The control variable of firm size has positive and significant effect on 4.0 IR adoption. As firms grow in size, both inter and intra administrative and operational responsibilities increases. Therefore, the adoption of certain types of technologies might become mandatory. Also, Firms size can facilitate new technology adoption, as larger organizations afford it has the capability to obtain required resources compared to smaller size firms. In contrast, firm age does not have a significant effect on 4.0 IR adoption. As many of newly established firms are open to new technologies compared to older firms who might resist changes and less motivated to adapt to new technologies. In the same time older firms might have better access to the market and its updates than smaller one, therefore they are exposed to technologies and more capable to obtain them.

5   Conclusion

This study enriches the logistics and 4.0 IR literatures by investigating the institutional factors' affecting logistics firms' intention to adopt 4.0 IR technologies in Oman, and the role of firm's characteristics on this effect. It discloses that both memetic and normative pressures strongly drive the adoption of 4.0 IR technologies in logistics sectors, whereas coercive pressure does not have any significant effect. The study also reveals that, in response to the ongoing institutional pressures, larger firms are more willing and able to adopt 4.0 IR technologies compared to smaller one. Whereas the firm's age is insignificantly related to 4.0 IR adoption. This study contributes to a better understanding of firms' drivers and decisions around the adoption of 4.0 IR technologies, especially in the context of Oman and its neighboring countries in the Gulf Corporate Council, a context which is going through huge logistics development but rarely been studied by the existing e-logistics studies. It is important for policymakers in Oman to formulate policies that drive and encourage further adoption of 4.0 IR technologies and create a technology roadmap to boost the logistics industry. The findings of this study may offer practitioners and policymakers guidelines for promoting the diffusion of 4.0 IR technology and suggest them to give further attention to the demands of all institutional forces in order to legitimate their operations.

Despite the theoretical and practical implications of this research, like many other empirical studies it suffers from some limitations which require further investigation by future researches. For example, our hypotheses were confirmed using data collected from a single country,

Oman. In fact, the context of Omani logistics industry, and its neighboring countries in the Gulf Corporate Council (GCC) have not been studied enough by previous empirical researches and, thus, we provided preliminary insight about the current status of adopting Logistics 4.0 in Oman, but Oman as a less developed country has different logistics expectations compared to other more developed countries. These differences in countries' logistics services expectations may influence organizational willingness to adopt various types of logistics innovation including 4.0 IR. Furthermore, in our study we could not control for the effects of firms' international orientation and ownership on their willingness to adopt Logistics 4.0, which may be considered as a moderating factor by future studies. Also, future studies may consider collecting data from more than one respondent per firm and consider using more objective measures for the constructs under investigation, which could be challenging because most of the time the innovation practices of logistics firms are not publicly available.

Acknowledgements The authors would like to thank the Technology Group in Oman Logistics Center (ASYAD) for sponging and supporting this research.

REFERENCES

Alavi, A., Nguyen, H.-O., Fei, J., & Sayareh, J. (2018). Port logistics integration: Challenges and approaches. *International Journal of Supply Chain Management, 7*(6), 389.

Armstrong, J. S., & Overton, T. S. (1977). Estimating nonresponse bias in mail surveys. *Journal of Marketing Research, 14*(3), 396–402.

Barratt, M., & Choi, T. (2007). Mandated RFID and institutional responses: Cases of decentralized business units. *Production and Operations Management, 16*(5), 569–585.

Barreto, L., Amaral, A., & Pereira, T. (2017). Industry 4.0 implications in logistics: An overview. *Procedia Manufacturing, 13*, 1245–1252. https://doi.org/10.1016/j.promfg.2017.09.045.

Chu, Z., Wang, L., & Lai, F. (2019). Customer pressure and green innovations at third party logistics providers in China: The moderation effect of organizational culture. *The International Journal of Logistics Management, 30*(1), 57–75.

Chu, Z., Xu, J., Lai, F., & Collins, B. J. (2018). Institutional theory and environmental pressures: The moderating effect of market uncertainty on innovation and firm performance. *IEEE Transactions on Engineering Management*, 65(3), 392–403.

DiMaggio, P. J., & Powell, W. W. (1983). The iron cage revisited: Institutional isomorphism and collective rationality in organizational fields. *American Sociological Review*, 48, 147–160.

Dubey, R., Gunasekaran, A., Childe, S. J., Blome, C., & Papadopoulos, T. (2019). Big data and predictive analytics and manufacturing performance: Integrating institutional theory, resource-based view and big data culture. *British Journal of Management*, 30(2), 341–361. https://doi.org/10.1111/1467-8551.12355.

Fornell, C., & Larcker, D. F. (1981). Evaluating structural equation models with unobservable variables and measurement error. *Journal of Marketing Research*, 18(1), 39–50.

Gupta, M., & George, J. F. (2016). Toward the development of a big data analytics capability. *Information & Management*, 53(8), 1049–1064.

Hair, J. F., Black, W. C., Babin, B. J., Anderson, R. E., & Tatham, R. L. (2010). *Multivariate data analysis* (Vol. 5). Pearson Education Inc.

Hofmann, E., & Rüsch, M. (2017). Industry 4.0 and the current status as well as future prospects on logistics. *Computers in Industry*, 89, 23–34. https://doi.org/10.1016/j.compind.2017.04.002.

Kauppi, K. (2013). Extending the use of institutional theory in operations and supply chain management research: Review and research suggestions. *International Journal of Operations & Production Management*, 33(10), 1318–1345.

Khaitan, S. K., & McCalley, J. D. (2015). Design techniques and applications of cyberphysical systems: A survey. *IEEE Systems Journal*, 9, 350–365.

Khalifa, M., & Davison, M. (2006). SME adoption of IT: The case of electronic trading systems. *IEEE Transactions on Engineering Management*, 53(2), 275–284.

Kuhlmann, A. S., & Klumpp, M. (2017). *Digitalization of logistics processes and the human perspective*. Paper presented at the Proceedings of the Hamburg International Conference of Logistics (HICL).

Liu, H., Ke, W., Wei, K. K., Gu, J., & Chen, H. (2010). The role of institutional pressures and organizational culture in the firm's intention to adopt internet-enabled supply chain management systems. *Journal of Operations Management*, 28(5), 372–384. https://doi.org/10.1016/j.jom.2009.11.010.

Maslarić, M., Nikoličić, S., & Mirčetić, D. (2016). Logistics response to the industry 4.0: The physical internet. *Open Engineering*, 6(1), 511–517.

Narayanan, S., Marucheck, A. S., & Handfield, R. B. (2009). Electronic data interchange: Research review and future directions. *Decision Sciences*, 40(1), 121–163.

Oloruntoba, R. (2018). Four theories for research in humanitarian logistics. In *The Palgrave handbook of humanitarian logistics and supply chain management* (pp. 675–712). Springer.

Parham, S., & Tamminga, H.-J. (2018). The adaptation of the logistic industry to the fourth industrial revolution: The role of human resource management. *Journal of Business Management & Social Science Research, 7*(9), 179–191.

Raitasuo, P., Finne, M., Kuula, M., & Ruiz-Torres, A. (2019). *Uncertainty, institutionalisation, and environmental performance in the logistics sector.* Paper presented at the Academy of Management Proceedings.

Scott, W. R. (2008). *Institutions and organizations: Ideas and interests.* Sage.

Sherer, S. A., Meyerhoefer, C. D., & Peng, L. (2016). Applying institutional theory to the adoption of electronic health records in the US. *Information & Management, 53*(5), 570–580.

Teo, H.-H., Wei, K. K., & Benbasat, I. (2003). Predicting intention to adopt interorganizational linkages: An institutional perspective. *MIS Quarterly, 27*(1), 19–49.

Wan, Q., Yuan, Y., & Lai, F. (2019). Disentangling the driving factors of logistics outsourcing: A configurational perspective. *Journal of Enterprise Information Management, 32*(6), 964–992.

Winkelhaus, S., & Grosse, E. H. (2019). Logistics 4.0: A systematic review towards a new logistics system. *International Journal of Production Research, 58,* 1–26. https://doi.org/10.1080/00207543.2019.1612964.

Zsidisin, G. A., Melnyk, S. A., & Ragatz, G. L. (2005). An institutional theory perspective of business continuity planning for purchasing and supply management. *International Journal of Production Research, 43*(16), 3401–3420.

Zucker, L. G. (1987). Institutional theories of organization. *Annual Review of Sociology, 13*(1), 443–464.

Industry 4.0: The Future of Manufacturing—Foundational Technologies, Adoption Challenges, and Future Research Directions

Suaad Jassem and Mohammad Rezaur Razzak

1 INTRODUCTION

The emergence of the new industrial era driven by advanced technology is enabling digitally controlled machines with embedded artificial intelligence (AI) to collect and analyze data from other devices, allowing for more agile, versatile, and reliable processes that result in higher-quality goods at lower costs (Madsen, 2019). Such transformational changes in the manufacturing environment are predicted to boost productivity, transform economic models, promote industrial development, and redeploy

S. Jassem (✉)
Department of Managerial and Financial Sciences, Al-Zahra College for Women, Muscat, Sultanate of Oman
e-mail: suaad@zcw.edu.om

M. R. Razzak
Department of Management, College of Economics and Political Science, Sultan Qaboos University, Muscat, Sultanate of Oman
e-mail: m.razzak@squ.edu.om

N. R. Al Mawali et al. (eds.), *Fourth Industrial Revolution and Business Dynamics*, https://doi.org/10.1007/978-981-16-3250-1_7

127

the workforce to higher skill tasks, allowing companies to compete in ways that were unimaginable in the past (Schwab, 2016). Labeled as the fourth industrial revolution or Industry 4.0 (I4.0), the term is currently one of the hottest buzzwords in the realm of business and industrial management around the world (Bessant, 2018). Since its' revelation at the Hannover Fair in 2011, it quickly attracted interest from academia and industry practitioners, not only in Germany, but all over the developed world (Rosa et al., 2019). The basic idea behind I4.0 is the application of advanced process management especially in manufacturing driven by emergent technologies such as digitalization, robotics, artificial intelligence, internet of things, additive manufacturing, etc.; completely transforming how products are made (Marr, 2018). I4.0 is also synony-mous with terms such as "smart manufacturing" , "integrated industry," etc. (Hofmann & Rusch, 2017). Industry pundits are looking at I4.0 as a key driver of the great digital transformation (Bessant, 2018), that is closely associated with other global trends such as augmented reality, cloud computing, big data analytics, circular business models, etc. (Marr, 2018).

Despite the popularity of the term Industry 4.0 in academic and management circles, the concept is difficult to articulate because there are more than 100 definitions of this term in engineering and manage-ment literature (Moeuf et al., 2018). According to the consulting firm; Mckinsey Group, I4.0 is a combination of many managerial and techno-logical concepts and is more or less a confluence of trends and promises of how products ought to be made by merging advanced technologies into the manufacturing environment (Baur & Wee, 2015). A more tech-jargon-loaded definition is: *"a new approach for controlling production processes by providing real-time synchronization of flows and by enabling the unitary and customized fabrication of products"* (Moeuf et al., 2018, p. 1118).

The idea of I4.0 is spreading rapidly and is becoming a broad approach encompassing much more than just conventional industrial production (Cotteleer, 2017). In fact, I4.0 has transformed from a specialized discourse to a widely talked about the phrase in vernacular of those connected to the phenomenon (Pfeiffer, 2017). The idea has driven the emergence of numerous other concepts such as Work 4.0, Supply Chain 4.0, Innovation 4.0, etc. (Reischauer, 2018), and is largely viewed as the future of manufacturing (Pfeiffer, 2017).

According to the GE Digital Report 2016, I4.0 transitions have the ability to generate value equal to 15–20% in efficiency gains (Schwab, 2016). This increased efficiency is partly attributed to reduction in system downtimes thanks to AI-driven remote sensing and preventive maintenance, as well as increased labor efficiency due to the automation of tasks that were previously accomplished manually. Moreover, the ability to analyze huge amount of data coming from industrial processes through sensors and actuators by machines and products connected to computing systems are a game-changer (Reischauer, 2018). The benefits of this new technology-enabled manufacturing ecosystem include reduction of inventories and the improvement of services such as shorter time-to-market, shorter lead times, fast delivery time, and lower freight costs with product quality that exceed customer expectations (Moeuf et al., 2018).

An ever-increasing number of companies are approaching the I4.0 paradigm, by connecting their factories, plants, machinery, and equipment through their intranet as well as to the Internet with the goal of improving efficiency and effectiveness (Rosa et al., 2019). In such hyper-connected industrial environments new challenges are emerging such as cybersecurity issues that represent one of the most pertinent concerns to be dealt with and is thus an integral part of I4.0.

1.1   Historical Overview: From Industry 1.0 to 4.0

The first industrial revolution (I1.0) is said to have started around 1760s and lasted till 1830s (Stearns, 2018). The core theme of I1.0 was the use of machines instead of human physical strength. These machines were mostly driven by coal-based steam-power and they improved both quantity and quality of products. The beginning came from the United Kingdom, and it soon spread to the United States and other parts of Europe (Stearns, 2018). Thereafter, with advancements made in engineering of machines, the second industrial revolution (I2.0) took roots in Europe, Japan, and USA from 1840s onwards. The main technological developments driving I2.0 came from generation of large-scale electricity and also improvements in transportation through railroads that eased the movement of raw materials and finished goods. Furthermore, iron and steel production became widespread leading to development of heavy industries.

Although the third industrial revolution or I3.0 emerged in the first half of the twentieth century where demand was fueled by two big

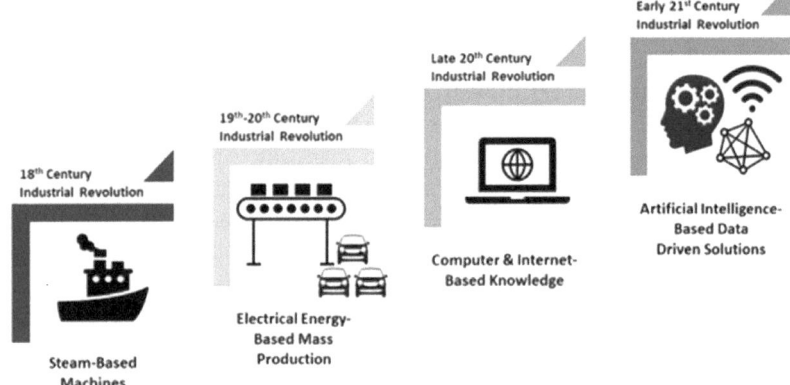

Fig. 1 Stages of Industrial Revolution (*Source* Complied from the Authors)

world wars, the major impetus came in the late 1960s with the advent of programmable logic controllers (Stearns, 2018). Breakthroughs in digital technologies developed in the 1970s solidified the foundation for I3.0, which was driven by information and communication technology. Furthermore, the development of information and communication technologies and range of machines driven by microprocessors from supercomputers to smaller devices with computational power that were increasing exponentially in their capacity to process data was transforming production processes to new heights of efficiencies and effectiveness never witnessed in the past. These digital transformations impacted everyday lives of consumers around the world.

With paradigm shift in technologies fueled by artificial intelligence, nanotechnology, cyber-physical space, and especially the seamless connectivity between machines and machines, and people and machines enabled by the Internet, the foundations of the fourth industrial revolution (I4.0) had been laid (Madsen, 2019).

The necessity of development and adoption of I4.0 was driven by multiple factors. Not only was there a phenomenal rise in the capacity to manufacture goods with great efficiency, but the myth of mutual exclusivity between volume and customization was shattered (Madsen, 2019). Now under the I4.0 ecosystem, manufacturers can produce customized

INDUSTRY 4.0: THE FUTURE OF MANUFACTURING ...   131

products in small to large quantities. Another area of concern for manufacturers in the advanced economies was the political pressure of not
exporting jobs to other low-cost destinations. By the implementation of
I4.0, the trend of jobs moving out of the industrialized economies to
developing countries is likely to decline as I4.0 is eliminating manual
tasks, and creating jobs in more high-skill areas such as programming and
engineering. The concept has come to prominence from publicity created
by the German industry in Hanover in 2011. Figure 1 shows the different
stages of the industrial eras.

2   Foundational Technology Driving Industry 4.0

The fourth wave of the industrial revolution driven by state-of-the-art
technological advancements are basically transforming the way goods and
services are produced. Despite the variation in the definitions of I4.0 in
the literature, there appears to be a consensus about nine technological
phenomena that are considered as the foundational building blocks of
Industry 4.0 in manufacturing (Ruessmann et al., 2015, see in Fig. 2).

In this digital transformation, the core concept is that subsystems
within a manufacturing system are connected through sensors, machines,
workpieces, and ICT systems along the entire production value chain
beyond a single enterprise (Stock & Seliger, 2016). These interconnected
systems, often referred to as cyber-physical systems, are able to collaborate
with each other using standard Internet-based protocols, analyzing data
to predict failure, re-configuring them as needed, and adjusting to any
changes. Combination of these technologies make it possible to gather
and analyze large quantity of data streaming into the system and across
machines, enabling faster and more flexible processes to produce high-
quality products at lower costs (Rosa et al., 2019). The I4.0 ecosystem
increases manufacturing productivity, enables rapid response to real-time
data, and modifies the workforce requirements, which ultimately makes
business organizations more competitive.

The nine (9) technology phenomenon that form the building blocks
of I4.0 along with their potential benefits for manufacturers are discussed
next.

Fig. 2 Nine technological phenomena Driving I4.0 (*Source* Compiled from the Authors)

2.1   Big Data Analytics

Big data analytics (BDA) powered by powerful computational capacity and sophisticated software enables the examination of large amounts of data to uncover hidden patterns and correlations that provide insights into the information. With today's technological achievements, it's possible to analyze gigabytes of data and derive meaningful and actionable information from it almost instantly that was a much slower and less efficient process with traditional business intelligence solutions (Rehman et al., 2019). Most organizations today realize that, by capturing all the data that stream into their businesses and by applying BDA they can derive significant information of value from it to take real-time decisions (Davenport & Dyche, 2013). Not so long ago, before the term "big data" was a familiar term, businesses were using basic analytics by observing numbers on spreadsheets that were examined manually to uncover insights and trends to help them take meaningful decisions. The quantum leap in analytical capabilities brought about by BDA, has made organizations more agile and providing them with a competitive edge never witnessed before (Rehman et al., 2019).

BDA supports organizations to harness the data streaming in and use them effectively to identify potential threats and opportunities. Such agility leads to smarter business moves, significantly more efficient operations, happier customers, and potentially higher profits. In the report by the International Institute of Analytics titled; *Big Data in Big Companies*, the authors interviewed more than 50 large business organizations to understand how they leverage information from BDA into their businesses (Davenport & Dyche, 2013). The report finds three common value-adding themes:

i. Cost Reduction: Technology platforms such as Hadoop and other pay-to-use cloud-based BDA systems provide significant cost advantages when it comes to storing huge amounts of data. Furthermore, they can also guide managerial decision-makers in identifying more efficient ways of doing business.

ii. Nimble Decision-making: The speed at which data is being processed into intelligent information, managers are now able to take real-time updates and analyze such information instantly—and make decisions or modify their current position. A good example of the benefits of nimble decision-making would be when a product quality issue is reported at the downstream of the value chain, it can be immediately addressed by other entities at the upper stream of the chain, thus saving large amounts of money by detecting and addressing problems as early as possible. For instance, at Infineon Technologies (a semiconductor manufacturer), by correlating single-chip data obtained in the testing phase at the end of the production process with data collected earlier in the process during the wafer stage, the company was able to reduce product failures. Firms like Infineon immensely benefit from such advance detection capabilities, which helps them discharge defective chips earlier in the manufacturing process and increase output quality (Rubmann et al. 2015).

iii. Rapid Deployment of New Products: With the ability to fathom customer perceptions in real-time, comes the power to respond to the market rapidly and give customers what they want. Davenport and Dyche (2013) point out that with the power of BDA, more companies are able to create new products to respond to customers' needs.

In the context of I4.0, the collection and comprehensive evaluation of large quantity of data from various sources (e.g., production equipment, customer-management systems, etc.) will become standard for organizational management to support real-time decision-making (Velasquez et al., 2018). In the past, BDA was more extensively used in customer behavior analysis related to marketing. However, big data analytics has lately found its way into the manufacturing environment, where it plays a vital role in improving production efficiency and quality, thus saving resources and improving equipment performance.

2.2   Autonomous Robots

Autonomous robots are smart devices that can perform complex functions in the physical world without the involvement of direct human supervision (Lima et al., 2019). Examples range from robot vacuum cleaners to drones and helicopters. The famous robotics scientist; Dr. George Bekey, surveyed implementation of robotic hardware in over 300 digital manufacturing systems, reviewing some of their areas of application, and studied the designs that drive these systems such as architectures, learning, control, manipulation, navigation, grasping, and mapping (Bekey, 2005). According to the aforesaid author, application of autonomous robotics is bound to become ubiquitous in the advanced economies, as governments are trying to limit export of industrial jobs to lower-cost destinations in developing countries. Robots will take over repetitive manual tasks and thus create demand for more skilled human labor force in engineering and programming (Lin et al., 2011).

Autonomous robots go hand-in-hand with artificial intelligence (i.e., machine learning) and are in a growing category of devices that includes unmanned drones (i.e., aerial robots) that can be self-programmed to perform tasks with minimum human intervention (Eimontaite et al., 2019). From robotics to flying vehicles with strong image and collecting data capabilities, they can vary considerably in capability, scale, dexterity, flexibility, and cost. Autonomous robots are increasingly being equipped with machine learning capabilities so that they can understand and learn from their environments and make decisions on their own (Lima et al., 2019).

Autonomous robotics is already fostering innovation into essential components of manufacturing such as the supply chain system and delivering significant value, primarily because they can:

- Improve accuracy and speed of routine operations, particularly in distribution, logistics, and warehousing;
- Enhance efficiency in processes through machines working in tandem with humans;
- Reduce the risk of employee injury and create safe work environments;
- Application of robots in manufacturing are also assisting companies in lowering long-term costs, balancing workforce utilization, increasing productivity of employees, reducing mistakes, reducing frequency of inventory checks, optimizing picking and sorting times, and gaining access to challenging or risky places.

Robotics and manufacturing are like a match made in heaven and are natural partners. Use of robots in the manufacturing landscape is gaining traction in most advanced industrial societies. Manufacturing processes driven by advanced automation have become a key part of manufacturing companies that are targeting competitive advantages through maximization of production efficiency and workplace safety. Use of robots in manufacturing enable execution of mundane repetitive tasks, bring down error rates to negligible levels, and free up human workers to spend more time and effort on more complex aspects of the manufacturing ecosystem such as programming and design. It is expected that soon robots will interact with other robots and work safely side by side with humans and learn from them (Lima et al., 2019). These robots will cost less and have a greater range of capabilities than those used in manufacturing today (Landscheidt et al., 2018).

Robots can be deployed in a variety of functions in manufacturing. For instance, fully autonomous robots are often used for performing work in elevated places, complex tasks that need and agility with precision (Kousi et al., 2019). Other manufacturing automation solutions include robots used to help people with more intricate tasks in handling parts and components including lifting, holding, and moving pieces around (Nair et al., 2019).

Automation in manufacturing helps businesses to compete globally by providing a cost-effective, feasible alternative to offshoring production and filling skills gaps in areas where human labor at competitive costs are difficult to find locally (Landscheidt et al., 2018). By utilizing robots in manufacturing, the human workforce can be re-deployed into functions that require creativity and innovation that call for a more complex neural network (i.e., the human mind), and are too complex for robots to perform currently. Such re-alignment in collaboration between man and machine are able to contribute to organizational growth and success. Increased efficiency, enhanced worker protection, and a superior bottom-line result (Nair et al., 2019).

2.2.1   *Efficiencies Created by Using Robotics in Manufacturing*

1. Robots applied in manufacturing create unparallel efficiencies starting from raw material handling to packaging and stacking of finished goods.
2. Robots may be programmed to operate 24 hours shifts through day and night if continuous production is required.
3. Autonomous robots are highly flexible and can be customized to perform extremely complex tasks.
4. Automation enabled by robots can be highly cost-effective for nearly every size of company, including small assembly operations.

2.2.2   *The Expanding Roles of Robotics in Manufacturing Environment*

The early generation of industrial robots were basically robotic arms that could move tooling through space along a defined and programmed path. They lacked the ability to go off from their defined paths and were not able to react to any unexpected obstacles making them confined to fixed pathways and structured environments (Gasparetto & Scalera, 2019). The limited use of early-stage robots in manufacturing, were typically applied to very high-volume assembly lines such as automobile body welding and painting. However, in the past decade with superior computational power along with emergence of artificial intelligence, significant advancements have been made in the utilization of industrial robots. Some examples of such advancements are:

- The advent of 2-D and 3-D vision systems that enable location of components in space has become a game-changer in industrial applications. These systems reduce the need to place parts and components in precise positions in the work envelope. As a result, the same robots are able to handle multiple products instead of just a single item (Muro & Andes, 2015).
- Force-sensing (i.e., pressure-sensing) allows control over side loads, enabling robotics applications such as trimming of molded or cast edges, precise assembly of components, etc. (Wilson & Daugherty, 2018).
- Delta robots (i.e., robot systems designed as parallelograms) are able to perform pick-and-place-type motions at high speeds that are ideal for applications such as packaging (Wilson & Daugherty, 2018).
- Cobots (collaborative robots) are designed to work safely along-side humans. They use sophisticated sensors eliminating the need to separate human workers from robots for safety. This enables humans and robots to work together on highly complex tasks that require precision assembly (Kildal et al., 2019).
- Self-driving vehicles (SDVs) are a more intelligent version of the Autonomous Guided Vehicles (AGVs) that were previously used in industrial centers. These self-driving vehicles use proprietary algorithms and advanced sensing technologies to navigate through production environments without any hindrance (Marks, 2019). The SDVs are mostly utilized in moving packages around production floors and large warehouses. These intelligent vehicles can be produced to work in a broad spectrum of sizes and functions. For example, in a warehouse setting they can be designed to integrate configurations and navigating capabilities that include lift items and then pick and drop off to other designated stations such as conveyors, cart engagement, or top plate of pallets.

Early-stage industrial robotics was mostly applied to high-volume, low-variety applications, but with new sensor technology and advanced software applications, robots are much faster and smarter and are able to work on more customized production ecosystems (Wilson & Daugherty, 2018).

2.3   *Simulation*

Simulation in the context of manufacturing systems refers to the application of software to develop computerized models of manufacturing systems, in order to analyze these systems and obtain important information such as impact of a local or specific action on the entire system (Machado et al., 2019). Computerized simulation is considered the second most popular field in management decision sciences among industrial engineers and manufacturing managers (Polenghi et al., 2018). However, its widespread is still evolving to keep pace with complex of some software packages that are being used by manufacturers these days.

These simulation techniques represent a valuable set of tools used in determining various aspects of managerial decisions. For example, they can be used to evaluate the potential benefits of investing in physical infrastructure and machinery in new production facilities, new warehousing systems, new distribution centers, etc. Simulation can also be used to forecast the performance of an existing or proposed new system and to compare between alternatives to take decisions (Polenghi et al., 2018). The most important goal of simulation in the field of manufacturing is the comprehension of the changes that occur on the total system as a result of some localized or specific change (Mourtzis et al., 2014). It is relatively easier to determine the differences caused by new adjustments in a localized system but it is relatively more challenging to assess the result of local changes on the whole system (Jahangirian et al., 2010). Simulation enables production managers to assess some measure of such consequences, for instance some measures that can be obtained by simulation are:

• Number of parts produced within a given period of time;
• Time required for processing each part in a process center;
• Time in the queue spent by each part;
• Time needed for transportation from one point to another;
• Deliveries made in time;
• Inventory build-up;
• In-process inventory;
• Utilization of machines and workers (percentage).

Some other benefits of simulation in manufacturing systems include calculating optimal resources required, validation of the proposed operational logic for controlling systems, data collected during modeling that may be used elsewhere, etc. (Komoto et al., 2019). Many manufacturers are now using simulations in manufacturing and assembly operations to take advantage of real-time data and utilize virtual models to mirror the physical world, which can include humans, machines, and products. This allows operators to optimize machine settings for the subsequent product in the virtual world prior to the physical changeover. This advantage brings down the machine setup time and enhances quality (Polenghi et al., 2018).

Simulation comprises an indispensable set of tools and methods for the successful implementation of the manufacturing ecosystem under I4.0, by allowing experimentation and validation of product, process, and system design and configuration (Polenghi et al., 2018). Therefore, the rapid adoption of digitalization in manufacturing in the context of the I4.0 has made simulation an essential part of the design and operation of manufacturing systems.

2.4 Horizontal and Vertical System Integration

I4.0 has further enhanced the importance of horizontal and vertical integration, making them an essential part of smart manufacturing systems. Horizontal integration in the context of I4.0 is the integration of value networks to enable collaboration between organizations in the value chain (Foidl & Felderer, 2015). Under this approach more than one company comes together through digital connectivity to cooperate and deliver superior products and services. This has become a common approach among German manufacturers of parts and components made for German automobile producers (Sindi & Roe, 2017). Vertical integration on the other hand is the integration of various hierarchical subsystems within the organization to create a flexible and reconfigurable production systems. The subsystems are typically connected through an enterprise resource planning/management platform that provides flexibility to reconfigure the manufacturing system as and when required (Foidl & Felderer, 2015). Such integration enables organizations to create customized products in a highly efficient manner.

With I4.0, the cyber-physical space has created more cohesive and synchronized interactions between companies and between departments or functions within companies, as universal data-integration networks evolve and enable truly automated value chains (Foidl & Felderer, 2015). Horizontal integration in the case of I4.0 envisions connected networks of cyber-physical and enterprise systems that function in tandem at unprecedented levels of automation and flexibility creating operational efficiencies along with capacity for high degrees of customization (Chukalov, 2017). Typically, horizontal integration in the I4.0 environment takes place at several subsystems within the manufacturing ecosystem of an organization. Some commonly observed areas of this phenomenon are:

i. On the Production Floor: As a result of uninterrupted machine connectivity, each production unit becomes a subsystem that is defined within the total production network. They are able to continuously update their performance status and autonomously respond to any dynamic feedback on production-related issues. The goal is to enable smart production systems to cost-effectively produce order quantities of different ranges and simultaneously reduce costly downtime through predictive maintenance (Chukalov, 2017).

ii. Across Multiple Production Facilities: When an organization has distributed production facilities, I4.0 advocates horizontal integration across plant levels through Manufacturing Execution Systems (MES) (Perez-Lara et al., 2018). In such a scenario, data on inventory levels, unexpected process delays, etc., are shared seamlessly across the entire organization. This enables the system to automatically shift production tasks among different facilities in order to meet targets and improve efficiency.

iii. Across the Entire Supply Chain: I4.0 also mandates data transparency and high levels of automated collaboration between the upstream and downstream entities on the supply and logistics chain (Perez-Lara et al., 2018). Third-party vendors and service providers to the organization are securely but closely incorporated horizontally into the enterprise's production and logistics control systems. A good example of such a collaborative network is the case of the Dell OptiPlex Plant in Round Rock, Texas, where all the supply chain partners are located within a pre-defined diameter from where the OptiPlex plant is situated. All the vendors are connected real-time with Dell's production mainframes and are sharing information between computers of Dell and the vendors continuously (Kumar & Craig, 2007).

Vertical integration in the case of I4.0 aims to tie together multiple functional layers within an organization from the production floor up through quality control, R&D, product management, sales and marketing, finance and accounting, human resource management, and so on. The objective being total transparency, data flows seamlessly up and down these layers so that both strategic and tactical decisions can be data-driven (Perez-Lara et al., 2018). The vertically integrated I4.0 manufacturer gains a crucial competitive edge by being able to respond with agility to changing market signals thus enabling rapid deployment of new products and services to the market place.

2.5 Industrial Internet of Things

Industrial Internet of Things (IIoT), also known as the Industrial Internet, a term coined by the General Electric Company (Gilchrist, 2016), was initially considered as a framework for industrial ecosystems in which numerous devices are connected with each other in a synchronized pattern through the application of advanced software-enabled computerized platforms in an ecosystem where machines communicate with other machines through the internet(Cotteleer, 2017). Nowadays, the IIoT has become increasingly more ubiquitous in the context of manufacturing ecosystems, as digitalization has become a strategic priority for many business organizations (Boyes et al., 2018).

Industrial Internet in manufacturing depicts an environment where machines, computers, and people are connected through the Internet and are enabling production operations using real-time business intelligence to bring about transformational outcomes for business organizations (Park, 2019). Thus, IIoT creates a synergy between intelligent machines and people at work with the help of advanced data analytics. Hence, it can be viewed as an ecosystem comprising of multiple industrial devices that are connected through networks resulting in systems that can collect and exchange information, then analyze and deliver valuable new insights from such inputs (Rehman et al., 2019). These insights drive smarter and faster business decisions for manufacturing companies. Simply stated, IIoT is transforming the manufacturing sector like never before by changing the way such organizations manage their daily operations.

The impact of IIoT is now applied even in many complex operations such as predictive analytics for functions such as corrosion inside industrial pipes, for providing real-time data from production centers to identify additional capacity in a plant, or accelerating development of new products by feeding service and operational data back to product design teams (Rehman et al., 2019). Industrial Internet powered by advanced software solutions, are giving organizations a tremendous edge. With the combination of machine-to-machine communication based on big data analytics, IIoT is driving unprecedented levels of efficiency, productivity, and performance (Olaf & Hanser, 2019). As a result, manufacturing companies in fields such as textile, original equipment manufacturing, pharmaceuticals, chemicals production, food and beverage industries, automobile manufacturing, and many other industries are experiencing unprecedented levels of operational and financial benefits (Rehman et al., 2019).

I4.0 means that more than ever, devices and embedded computing will go hand-in-hand. This has allowed field devices (e.g., equipment on the production floor) to interact with both; equipment and centralized controllers. It also has automated analysis and decision-making, allowing it to react in real-time (Olaf & Hanser, 2019).

2.6   Cybersecurity

With the immense progress brought to manufacturing ecosystems through hyperconnected technology applications, comes the specter of cyber threats due to unauthorized infiltration into the system. Therefore, in the context of I4.0, cybersecurity plays a vital role in preventing the loss of control over the system, and data integrity, and thus organizational competitiveness. In fact, critical industrial equipment may be at risk to cyber-attacks from unidentifiable sources, which have the potential to cause damage to entire business operations. According to Cisco's 2018 Annual Cybersecurity Report, 31% of industrial organizations have experienced cyber-attacks on their operational systems; while, 38% expect attacks to extend into their information communication networks to their operational systems. Although 75% of cybersecurity experts involved with manufacturing organizations perceive cyber-attacks as a clear and present danger, less than 16% of such organizations appear to be sufficiently prepared to face these potential threats (Ervural & Ervural, 2018). Such vulnerabilities are mostly attributed to the lack of unified standards for organizations to benchmark against. This is coupled with a severe shortage of managerial and technical skills necessary to implement them (Dawson, 2018).

Organizations based in the European Union and North America have been prioritizing moves in the direction of standardizing cybersecurity protocols. For instance, in 2017, European Cyber Security Organization (ESCO) put together a large volume of documents about existing standards and specifications connected to Cybersecurity with particular focus on the European Digital Single Market (Pereira et al., 2017). The above document helps to determine the existing protocols used by companies to address the cybersecurity challenges and potential vulnerable areas. In addition to the above, the International Electrotechnical Commission (IEC) based in Geneva, Switzerland has released broad guidelines on security of information and privacy of data, that looks at important guidelines to be covered in subsequent IEC publications and elaborates on how they may be effectively implemented (Hussain et al., 2019). I4.0 necessitates increased connectivity; as a result there arises an imminent requirement for the protection of critical infrastructure and industrial systems from cybersecurity threats. Therefore, a reliable and secure communication system as well as sophisticated identity and access management of machines for users is vital (Pereira et al., 2017).

The I4.0 has a very close association with cybersecurity. In recent times, a number of I4.0 cybersecurity incidents have been reported, additionally stressing the need to strengthen protection from cyber-attacks (Dawson, 2018). Industrial operators who utilize IoT and I4.0 solutions are particularly vulnerable to such threats. The need to improve cybersecurity under the I4.0 protocol has become crucial, since the possible impact of cyber threats range from concerns regarding physical security to downtimes in production, damage to products and equipment, as well as potential for substantial reputational and financial losses (Dawson, 2018).

Steve Purser, head of Core Operations Department at ENISA, said: *"The advanced digitalization envisaged within the Industry 4.0 framework is a paradigm shift in the way industries operate and blurs the boundaries between the physical and digital world. With a great impact on citizens' safety, security and privacy due to its cyber-physical nature, the security challenges concerning Industry 4.0 and IoT are significant. Today, ENISA publishes the study that addresses those challenges and, combined with the baseline IoT security work lays the foundations for a secure industrial IoT ecosystem. IoT, together with Industry 4.0 cybersecurity are the springboard for a safer and more resilient connected world."* (ENISA, 2018).

Any manufacturer that uses IoT systems is directly or indirectly affected by the possibility of cyber threats. Malicious attacks, especially with regard to large organizations, can result in significant financial losses as well as incalculable losses through data corruption, triggering system failures, privacy violations, loss of credibility, consumer frustration, and overall market losses. Most large companies have considerably strengthened their cybersecurity capabilities in recent years (Hussain et al., 2019). To limit the risk of cyber-attacks, millions of dollars have been spent on developing new techniques with technological investments in IT security. If the application of Industrial Internet of Things (IIoT) keeps growing at the current pace without corresponding cybersecurity measures, the modern manufacturing ecosystem will become a more desirable target for cyber-attackers (Dawson, 2018).

2.7   The Cloud

Cloud computing is basically on-demand delivery of information technology resources, made available over the Internet by other providers. Hence, organizations are able to avail services of powerful servers, large data storage spaces, robust databases, and software applications without actually owning the physical hardware or the software. Cloud technology is a crucial enabler of the I4.0. As I4.0 gains popularity due to developments such as the IoT, automation and robotics, artificial intelligence, etc., organizations would not be able to cope without the support of the cloud to effectively free these technologies from the connectivity restraints of on-site servers (Velasquez et al., 2018).

Business organizations that are leveraging the benefits of cloud computing in manufacturing processes can operate more efficiently and focus on their core competencies by freeing the management from the burden of maintaining their own IT infrastructure. Hence, with the support of the cloud, management can simply utilize the software and hardware resources based on their needs, enabling them to scale up or scale down whenever required (Velasquez et al., 2018). By using cloud technology support from external providers, business organizations can avoid major capital investments in computer hardware and software licenses.

In the case of I4.0 the advantages of advanced cloud technology goes beyond reliability, scalability, and storage and the associated cost savings. Cloud technology has provided a platform for companies especially start-up ventures, to create a variety of innovations without substantial investment in IT infrastructure, thus enabling them to evolve rapidly and become significant players in the market. In fact, Harvard University research indicates that small firms and start-up ventures are able to leverage the benefits of cloud technology and compete with large established companies by avoiding capital expenditures in IT infrastructures (Bloom & Pierri, 2018). The cloud has increased borderless collaboration among sites and firms in sharing data and cooperating in production-related activities. Simultaneously, cloud technology can increase their performance, allowing for reaction times of only a few milliseconds. The current trend is that device application programs are rapidly migrating to the cloud, allowing production processes to benefit from more data-driven services (Hardy, 2018).

2.8   Additive Manufacturing

The term "additive manufacturing" is synonymously used with the term "3-D printing" which basically refers to technologies that develop three-dimensional objects one superfine layer at a time. Each successive layer is bonded to the preceding layer of material, which is typically partly or fully melted on to each layer (Mehrpouya et al., 2019). Different materials, such as thermoplastics, metal powder, ceramics, composites, glass, and even edible material, may be used to layer material (D'Aveni, 2015). Software is used to create digital designs that "slice" the physical object into super-thin layers.

This digital information guides the path of a nozzle that precisely deposits material upon the preceding layer.

Additive manufacturing processes 3-D parts directly from CAD (Computer-Aided Design) models by adding a layer of material upon another and is far more efficient than the traditional subtractive manufacturing process (Mehrpouya et al., 2019). The application of this technology has been adopted by numerous industrial organizations especially in the field of aerospace, automobile manufacturing, biomedical fields, etc., it allows manufacturers to produce small to large quantities of goods with customized specifications, thus enabling mass customization.

The association of additive manufacturing with I4.0 is that companies can create scalable factories built to meet the challenges of a globalized marketplace (D'Aveni, 2018). By enabling companies to rapidly produce and bring newly customized products to the market, such agility has profoundly impacted the capabilities of manufacturers. Manufacturing lead times can be reduced substantially, hence new product designs will require substantially shorter time-frames for them to reach customers, and capacity can be adjusted based on customer demand (Mehrpouya et al., 2019). It is expected that with I4.0, additive manufacturing approaches will become widespread in producing small batches of customized products that offer development and production advantages that can adjust to complexities of the design.

2.9   Augmented Reality

Augmented reality (AR) integrates and projects information generated by computers into the physical world. At present majority of the Augmented reality applications combine computer graphics into the view of the user's surroundings. These AR applications are providing capability to enhance the user-experience of system applications where such users are able to access and interact with information that has a spatial relation to environment surrounding them (Cipresso et al., 2018).

In the manufacturing ecosystem, industrial augmented reality has become an integral component of I4.0, as it enables workers to access digital information and overlay that information with their physical world (Flavian et al., 2019). Although the adoption of this technology across all industry type is still in the early stages, the annual growth rate of the industrial AR market is predicted to grow rapidly (Iftene & Trandabăt, 2018).

The application of AR is an important component of I4.0 (Cipresso et al., 2018). Within the I4.0 environment, industrial AR promotes a human-centric industrial environment where the technology is used to augment the workers' integrated view of the physical and computer-generated worlds. This advanced technological development has enabled people to access the digital world through layers of information that are positioned on top of the physical world (Flavian et al., 2019). Therefore, AR integrates physical and the virtual reality, which means it augments the real world without replacing it.

The technology has a wide range of applications in the manufacturing environment. For instance, AR can be used to design and produce applications, utilized in assembly operations, deployed for training and online guidance systems for operators, etc. (Iftene & Trandabăt, 2018). Additionally, it is used in warehousing and logistics with applications such as "pick-by-vision" where AR indicates quantities and locations of parts and sends repair instructions over mobile devices. It also has wide use in quality assurance, ensuring workers' safety, etc. These applications of AR in the context of I4.0 are currently in their early stages, but soon, companies are expected to make much broader use of AR to provide workers with real-time information to improve decision-making and simplify work procedures (Flavian et al., 2019).

3   CHALLENGES IN ADOPTION OF INDUSTRY 4.0

The apparent benefits of I4.0 are more or less recognized in the advanced industrial societies, however, the repercussions of massive losses in human jobs as a result of smart factories is still a matter of concern for policy makers (Muro & Andes, 2015). However, the overall notion of human tasks being completely replaced by autonomous and hyperconnected machines has been addressed by many research studies that suggest that increase in efficiency in producing goods and services will free up human capacity to focus on high-skill activities that can't be yet be done by machines. Instead of concern for replacement of manual jobs done by humans, the workforce will be deployed in more high-skill areas plus autonomous machines will be working side by side with humans in a safer manner (Wilson & Daugherty, 2018). Despite the recognition of the benefits of I4.0 there are several challenges to adoption of the fourth industrial revolution.

The fact that even highly industrialized societies such as the USA are approaching I4.0 cautiously is evident from the fact that at present only seven (7) countries are at the forefront of wholesale adoption of I4.0, these are: Germany, Canada, Japan, China, Australia, Austria, and Switzerland (Tassel, 2019). According to a report by the World Economic Forum in 2019, three of the main challenges that come in the way of adoption of I4.0 are: (i) lack of unified leadership required to build a regional ecosystem to support all elements needed to foster adoption of I4.0, (ii) lack of sufficient availability of managerial and engineering talent to pioneer the transition to I4.0, and (iii) data ownership concerns especially when external entities such as vendors are part of the system (Tassel, 2019).

3.1   Building an Ecosystem Through Private–Public Partnerships

In order for companies to make the transition to a fully implemented I4.0 environment, there has to be unified leadership that comprises cooperation between policy makers, industry, and academia (Renjen, 2018). The Germans have been at the forefront of adopting I4.0 mainly because of the unified leadership at both national and regional levels that brings the three main stakeholders together (Tassel, 2019). The nations that are making a relatively smooth transition into I4.0 have close cooperation between national and regional organizations in government, research institutions under universities, and business organizations (Renjen, 2018). Furthermore, these countries all have advanced engineering capabilities and are taking lead in innovation in merging the manufacturing workplace with the cyber-physical space. Therefore, for other countries to make such a transition to I4.0, similar unified leadership is warranted through private–public partnerships.

3.2   Creating Talent Pool for I4.0 Through Industry-Academia Collaboration

One of the key features of societies that are making a successful journey toward the new industrial era are that they are generating a sufficiently large pool of talented managers, critical thinkers, engineers, programmers, and technicians who can work toward developing and maintaining SMART (Specific, measurable, achievable, realistic, and time efficient) factories (Gann et al., 2018). Such nations are leveraging the synergies created by academia-industry collaboration for paving the path forward to overcoming hurdles related to creating the workforce of tomorrow (Hou et al., 2019).

Through partnerships between universities and manufacturing companies the entire curricula or a substantial portion of the coursework for students along with industry immersion programs can be revamped to prepare the new graduates from universities and Polytechnique institutes to be oriented toward the I4.0 environment. Other than a small cluster of nations such as Canada, Germany, Japan, China, Switzerland, etc., the curricula at universities in management and engineering have not been updated to enable development of the necessary workforce to shoulder the responsibility of supporting smart hyperconnected factories (Renjen, 2018). Furthermore, the research outputs from universities related to I4.0

appears to be concentrated in a few countries, meaning that most other societies are not undertaking sufficient research work and thus developing enough teachers and trainers to build the workforce required for I4.0. Figure 3 shows the country-wise output of research papers published by universities related to I4.0.

The information shown in the above figure indicates highly skewed data in favor of just a few nations at the forefront of I4.0 in terms of published research. Therefore, without sufficient collaboration between institutions of higher learning and industry, the crisis for a workforce with industry-ready skills to support I4.0 will remain a hurdle.

Despite the challenges, there appear to be several initiatives that are paving the path forward toward successful models of academia-industry partnerships. A report published by the consulting firm, Deloitte, that is titled *Future of Manufacturing: Making Things in a Changing World*, elaborates on how the digital transformation is being accelerated by the global companies. For instance, Siemens and IBM are working closely with research institutes housed within universities around the world, to

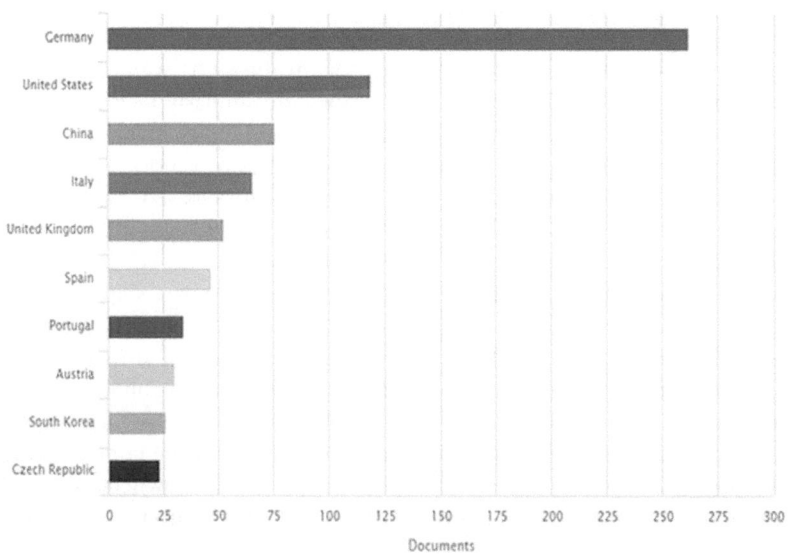

Fig. 3 Research papers published on I4.0 country-wise (*Source* World Economic Forum Report, 2019)

bring high-tech manufacturing to industries that previously were not considered for such changes yet (Hagel et al., 2015).

An example of such an industry is the pharmaceutical industry. For instance, the University College Dublin along with the Technological University Dublin, are working with pharmaceutical manufacturing companies such as Novartis, Lonza, Pfizer, and AstraZeneca. The outcome of such partnership is deployment of autonomous robots in production of medicine for drugs that need extreme conditions, such as the MRNA-type COVID-19 vaccine produced by Pfizer and Moderna, which requires temperatures around -94 degrees Fahrenheit at all times. This collaboration has now borne fruit, as their experiences and innovations are now being used in the production of COVID-19 vaccines at speeds that were previously never achieved in the pharmaceutical industry (Lurie et al., 2020).

3.3   Data Ownership Concerns

The nature of the manner in which manufacturers work in the I4.0 environment requires close coordination with third-party vendors that enables data sharing between the factory and the vendors. For instance, the Dell OptiPlex plant in Texas allows sharing of inventory as well as demand data for parts with their component vendors and similarly Dell can access the database of the vendors to check the availability of inventory. Such visibility enables Dell to automatically route orders to vendors with sufficient stocks, while vendors can plan their inventory based on variation in demand and stocks at Dell OptiPlex (Kumar & Craig, 2007). Such highly collaborative supply chain networks with complete visibility of each other's sensitive data may also induce concerns for ownership of information between the buyer and the seller (Agostini & Filippini, 2019). Organizations such as Dell have been extremely successful at maintaining their close collaboration without any major disruptions mainly due to the fact that one of the organizations (i.e., in this case; Dell Computers) takes the lead in codifying the rules of engagement between the company and vendors (Min et al., 2019). Hence, the concerns for data ownership in the I4.0 ecosystem needs, to be codified through private–public partnership efforts.

The issue concerning ownership of data needs to be addressed at the earliest through a global body comprising of majority of the nations. This point is highlighted by a report in GE Digital, concerning future litigations that are likely to emerge over data ownership. Such litigations are likely to be more complex since data is digitally stored in remote servers in several storage centers around the world. Hence even if claims arise, there will be a question on which court has jurisdiction to rule over any such claims (Hitchin, 2017).

4 FUTURE RESEARCH DIRECTIONS

Among the many obstacles faced in dealing with the issue of I4.0, is the fact that there are numerous definitions of I4.0 spread throughout the literature (Moeuf et al., 2018). Therefore, in order to bring academic research on I4.0 to a more unified platform, there is a need for concerted effort to bring about a confluence of these definitions. This may be a useful endeavor for future academic research in collaboration with industry thinktanks and policymakers.

The transition to I4.0 has become a desired goal for many nations including those are still developing, but it remains an elusive dream for most. The challenges of adoption of I4.0 in general highlighted by the World Economic Forum Report (mentioned in Sect. 3 above) presents three generic obstacles faced by societies in building an I4.0 ecosystem. The hurdles need to be assessed based on localized context. For instance, many countries in the GCC region are planning to get out of reliance on primary mineral resources as their source of economic sustenance and move on to a more diversified economy through industrial development. A good example of such an initiative is the Saudi Vision 2030, under which the present government has made diversification of the economy their primary goal. This has led to a massive and highly ambitious project under the name of NEOM City at the Western region of Saudi Arabia (Camburn, 2020). Such initiatives would make sense considering the fact that many such nations have relatively low local populations and high wages. Therefore, research focus needs to be emphasized on local context of each of the GCC countries to identify the obstacles that need to be addressed in order to adopt I4.0 and radically transform their economies.

The issue of academia and industry collaboration is yet to be sufficiently explored in most parts of the world. Considering the fact that, countries such as Germany, Canada, Japan, and China have made admirable strides in such collaborative projects, their models need to be studied closely and adjusted to the local context through mutually beneficial partnerships between business organizations and universities with the support of government. Therefore, research work in such countries needs to be focused on developing suitable models of collaboration between academia and industry.

5   CONCLUSION

This book chapter presents a narrative about I4.0 in the realm of manufacturing. The first part delves into describing the nine (9) foundational technological phenomena that form the backbone of I4.0. The objective behind presenting the core technology was due to the fact that the concept of I4.0 is very difficult to articulate, mainly due to the variation in the definitions of I4.0 in the literature. The second part covers discussions on the three primary challenges faced by societies in adopting I4.0 based on a study published by the World Economic Forum in 2019. This study acknowledges the challenges and also proposes potential solutions to overcome these generic hurdles. The last part of the study focuses on a few research agendas for future research undertakings that are not all inclusive by any means, as the topic of I4.0 is vast. However, focus is laid down on contextual research that must be done to identify how other nations that want to diversify their economies can take part in I4.0. It is expected that the outcome of this study will be useful to academic researchers as well as to industry practitioners.

The current focus of research on I4.0 is likely to remain heavily skewed in favor of qualitative research for some time. The reason for this is that for empirical studies to be valid and meaningful, there has to be sufficient opportunities for data collection from a large enough population for drawing any meaningful inferences. However, it must be mentioned that in order for theoretical developments related to the field of usage of I4.0 technologies, there is a possibility that the development of this field may follow the same pattern that was observed in the field of "technology acceptance behavior" that was actually ignited by the field of information systems due to proliferation of the use of computers (e.g., Davis, 1989; Venkatesh et al., 2012).

Studies on Information and communication technology (ICT) usage behavior lead to the development of some of the most widely applied theoretical models such as TAM (Technology Acceptance Model), TAM-2, UTAUT (Unified Technology Acceptance and Use Model), UTAUT-2, meta-UTAUT, etc. (Dwivedi et al., 2019). These models had a profound impact on understanding of relationships between technology and human perceptions about technology. Similarly, as the field of I4.0 matures both in theory and practice, there are likely to emerge theoretical models that will be able to predict how various constructs relate to each other in this field. Such initiatives will lead to further understanding of this phenomenon, that making is yet trying to come to terms with.

REFERENCES

Agostini, L., & Filippini, R. (2019). Organizational and managerial challenges in the path toward Industry 4.0. *European Journal of Innovation Management*, *22*(3), 406–421.

Baur, C., & Wee, D. (2015 June). Manufacturing's next act. *McKinsey Quarterly* (online). Available at https://www.mckinsey.com/business-functions/operat ions/our-insights/manufacturings-next-act.

Bessant, J. (2018). The great transformation: History for a techno-human future.

Bekey, G. A. (2005). *Autonomous Robots*. MIT Press.

Bloom, N., & Pierri, N. (2018). Research: Cloud computing is helping smaller, newer firms [17] Compete. *Harvard Business Review*.

Boyes, H., Hallaq, B., Cunningham, J., & Watson, T. (2018). The industrial internet of things (IIoT): An analysis framework. *Computers in Industry, 101*, 1–12.

Camburn, M. (2020, November). Saudi Arabia's Vision 2030 has had a significant impact through RETT and VAT changes. *Forbes* (online). Available at https://www.forbesmiddleeast.com/leadership/opinion/saudi-arabias-vis ion-2030-has-had-a-significant-impact-through-rett-and-vat-changes-1.

Chukalov, K. (2017). Horizontal and vertical integration, as a requirement for cyber-physical systems in the context of Industry 4.0. *International Scientific Journal "Industry 4.0"*, *2*(4), 155–157.

Cipresso, P., Giglioli, I. A. C., Raya, M. A., & Rive, G. (2018). The past, present, and future of virtual and augmented reality research: A network and cluster analysis of the literature. *Frontier in Psychology, 9*, 9–20.

Cotteleer, M. (2017, April). How industrial IoT will reshape manufacturing—And the bottom line. *Harvard Business Review*. Available at https://hbr.org/webinar/2017/04/how-industrial-iot-will-reshape-manufacturing-and-the-bottom-line.

D'Aveni, R. (2015). The 3-D printing revolution. *Harvard Business Review*, *93*(5), 40–48.

D'Aveni, R. A. (2018). The 3 D printing playbook. *Harvard Business Review,* *96*(4), 106–113.

Davenport, T. H., & Dyche, J. (2013). Big data in big companies. International Institute for Analytics, pp.1–33.

Davis, F. D. (1989). Perceived usefulness, perceived ease of use, and user acceptance of information technology. *MIS Quarterly, 13*(3), 319–340.

Dawson, M. (2018). Cyber security in Industry 4.0: The pitfalls of having hyperconnected systems. *Journal of Strategic Management Studies, 10*(1), 19–28.

Dwivedi, Y. K., Rana, N. P., Jeyaraj, A., Clement, M., & Williams, M. D. (2019). Re-examining the Unified Theory of Acceptance and Use of Technology (UTAUT): Towards a revised theoretical model. *Information System Frontier, 21*, 719–734.

Eimontaite, I., Gwilt, I., Cameron, D., Aitken, J. M., Rolph, J., Mokaram, S., & Law, J. (2019). Language-free graphical signage improves human performance and reduces anxiety when working collaboratively with robots. *The International Journal of Advanced Manufacturing Technology, 100*(4), 55–73.

ENISA, (2018, November). Cybersecurity is a key enabler for Industry 4.0 adoption (online). Available at https://www.enisa.europa.eu/news/enisa-news/cybersecurity-is-a-key-enabler-for-industry-4-0-adoption.

Ervural, B. C., & Ervural, B. (2018). Overview of cyber security in the industry 4.0 Era. In *Industry 4.0: Managing the digital transformation* (pp. 267–284). Cham: Springer.

Flavian, C., Ibanez-Sanchez, S., & Orus, C. (2019). The impact of virtual, augmented and mixed reality technologies on the customer experience. *Journal of Business Research, 100*, 547–560.

Foidl, H., & Felderer, M. (2015, November). Research challenges of industry 4.0 for quality management. In *International Conference on Enterprise Resource Planning Systems* (pp. 121–137). Cham: Springer.

Gann, D., Montressor, F., & Eisenberg, J. (2018, November). *3 ways to nuture collaboration between universities and industry.* World Economic Forum Report. Available at https://www.weforum.org/agenda/2018/11/3-ways-to-nurture-collaboration-between-universities-and-industry/.

Gasparetto, A., & Scalera, L. (2019). A Brief history of industrial robotics in the 20th century. *Advances in Historical Studies, 8*(1), 24–35.

Gilchrist, A. (2016). *Industry 4.0: The industrial Internet of things.* New York: Apress.

Hagel III, J., Brown, J. S., Kulasooriya, D., Giffi, C., & Chen, M. (2015). *The future of manufacturing-making things in a changing world.* Future Bus (online). Available at The future of manufacturing | Deloitte Insights.

Hardy, Q. (2018, February). How cloud computing is changing management. *Harvard Business Review* (online). Available at https://hbr.org/2018/02/how-cloud-computing-is-changing-management.

Hitchin, P. (2017). *The industrial Internet of things: Who really owns your data?* GE Digital Report (online). Available at https://www.ge.com/power/transf orm/article.transform.articles.2017.may.the-industrial-internet-of-thi.

Hofmann, E., & Rusch, M. (2017). Industry 4.0 and the current status as well as future prospects on logistics. *Computers in Industry, 89*, 23–34.

Hou, B., Hong, J., Wang, H., & Zhou, C. (2019). Academia-industry collaboration, government funding and innovation efficiency in Chinese industrial enterprises. *Technology Analysis & Strategic Management, 31*(6), 692–706.

Hussain, S. S., Ustun, T. S., & Kalam, A. (2019). *A review of IEC 62351 security mechanisms for IEC 61850 message exchanges.* IEEE Transactions on Industrial Informatics.

Iftene, A., & Trandabăţ, D. (2018). Enhancing the attractiveness of learning through augmented reality. *Procedia Computer Science, 126*, 166–175.

Jahangirian, M., Edabi, T., Naseer, A., Stergioulas, L. K., & Young, T. (2010). Simulation in manufacturing and business: A review. *European Journal of Operational Research, 203*(1), 1–13.

Kildal, J., Martin, M., Ibon, I., & Maurtua, I. (2019). Empowering assembly workers with cognitive disabilities by working with collaborative robots: A study to capture design requirements. *Procedia CIRP, 81*, 797–802.

Komoto, H., Kondoh, S., Furukawa, Y., & Sawada, H. (2019). A simulation framework to analyze information flows in a smart factory with focus on run-time adaptability of machine tools. *Procedia CIRP, 81*, 334–339.

Kousi, N., Stoubos, C., Gkournelos, C., Michalos, G., & Makris, S. (2019). Enabling Human Robot Interaction in flexible robotic assembly lines: An Augmented Reality based software suite. *Procedia CIRP, 81*, 1429–1434.

Kumar, S., & Craig, S. (2007). Dell, Inc'.s closed loop supply chain computer assembly plants. *Information Knowledge Systems Management, 6*, 197–214.

Landscheidt, S., Kans, M., Winroth, M., & Wester, H. (2018). The future of industrial robot business: Product or performance based? *Procedia Manufacturing, 25*, 495–502.

Lima, F., de Carvalho, C. N., Acardi, M. B., dos Santos, E. G., de Miranda, G. B., Maia, R. F., & Massote, A. A. (2019). Digital manufacturing tools in the simulation of collaborative robots: Towards Industry 4.0. *Brazilian Journal of Operations & Production Management, 16*(2), 261–280.

Lin, P., Abney, K., & Bekey, G. (2011). Robot ethics: Mapping the issues for a mechanized world. *Artificial Intelligence, 175*(5–6), 942–949.

Lurie, N., Saville, M., Hatchett, R., & Halton, J. (2020). Developing Covid-19 vaccines at pandemic speed. *The New England Journal of Medicine, 382*, 1969–1973.

Machado, R. H. C., Helleno, A. L., de Oliveira, M. C., Santos, M. S. C. D., & Dias, R. M. D. C. (2019). Analysis of the influence of standard time variability on the reliability of the simulation of assembly operations in manufacturing systems. *Human Factors, 61*(4), 627–641.

Madsen, D.O. (2019). The emergence and rise of Industry 4.0 viewed through the lens of management fashion theory. *Administrative Sciences, 9*(3), 1–25.

Marks, M. (2019). *Robots in space: Sharing our world with autonomous delivery vehicles*. Available at SSRN 3347466.

Marr, B. (2018, September 2). What is Industry 4.0? Here's a super easy explanation for anyone. *Forbes* (online). Available at https://www.forbes.com/sites/bernardmarr/2018/09/02/what-is-industry-4-0-heres-a-super-easy-explanation-for-anyone/#79db95459788.

Mehrpouya, M., Dehghanghadikolaei, A., Fotovvati, B., Vosooghnia, A., Emamian, S. S., & Gisario, A. (2019). The potential of additive manufacturing in the smart factory Industrial 4.0: Review. *Applied Sciences, 9*, 2–34.

Min, S., Zacharia, Z. G., & Smith, C. D. (2019). Defining supply chain management: In the past, present, and future. *Journal of Business Logistics, 40*, 44–55.

Moeuf, A., Pellerin, R., Lamouri, S., Tamayo-Giraldo, S., & Barbaray, R. (2018). The industrial management of SMEs in the era of Industry 4.0. *International Journal of Production Research, 56*, 1118–1136.

Mourtzis, D., Doukas, M., & Bernidaki, D. (2014). Simulation in manufacturing: Review and challenges. *Procedia CIRP, 25*, 213–229.

Muro, M., & Andes, S. (2015, June). Robots seem to be improving productivity, not costing jobs. *Harvard Business Review* (online). Available at https://hbr.org/2015/06/robots-seem-to-be-improving-productivity-not-costing-jobs.

Nair, V. V., Kuhn, D., & Hummel, V. (2019). Development of an easy teaching and simulation solution for an autonomous mobile robot system. *Procedia Manufacturing, 31*, 270–276.

Olaf, J. M., & Hanser, E. (2019). Manufacturing in times of digital business and Industry 4.0-The industrial Internet of things not only changes the world of manufacturing. In *Advances in manufacturing engineering and materials* (pp. 11–17). Cham: Springer.

Park, J. H. (2019). Advances in future Internet and industrial Internet of things. *Symmetry, 11*(2), 244.

Pereira, T., Barreto, L., & Amaral, A. (2017). Network and information security challenges within Industry 4.0 paradigm. *Procedia Manufacturing, 13*, 1253–1260.

Perez-Lara, M., Saucedo-Martinez, J.A., Marmolejo-Saucedo, J.A., Salis-Fierro, T.E., & Vasant, P. (2018, November). *Vertical and horizontal integration systems in Industry 4.0*. Wireless Networks (online). Available at https://www.researchgate.net/publication/329111116_Vertical_and_horizontal_integration_systems_in_Industry_40.

Pfeiffer, S. (2017). The vision of "Industrie 4.0" in the making—A case of future told, tamed, and traded. *Nanoethics, 11*(1), 107–121.

Polenghi, A., Fumagalli, L., & Roda, I. (2018). Role of simulation in industrial engineering: Focus on manufacturing systems. *IFAC-PapersOnLine, 51*(11), 496–501.

Rehman, M. H., Yaqoob, I., Salah, K., Imran, M., Jayaraman, P. P., & Perera, C. (2019). The role of big data analytics in industrial Internet of things. *Future Generation Computer Systems, 99*, 247–259.

Reischauer, G. (2018). Industry 4.0 as policy-driven discourse to institutionalize innovation systems in manufacturing. *Technological Forecasting and Social Change, 132*, 26–33.

Renjen, P. (2018). Industry 4.0. Are you ready? *Deloitte Rev, 22*, 9–11.

Rosa, P., Sassanelli, C., Urbinati, A, Chiaroni, D., & Terzi, S. (2019). Assessing relations between Circular Economy and Industry 4.0: a systematic literature review. *International Journal of Production Research* (in press). Available at https://www.tandfonline.com/doi/full/https://doi.org/10.1080/00207543.2019.1680896.

Ruessmann, M., Lorenz, M., Gerbert, P., Waldner, M., Justus, J., Engel, P., & Harnisch, M. (2015). Industry 4.0: The future of productivity and growth in manufacturing industries. *Boston Consulting Group, 9*(1), 54–89.

Schwab, K. (2016). *The 4th Industrial Revolution: What it means, how to respond.* GE Digital Reports 2016. Available at https://www.weforum.org/about/the-fourth-industrial-revolution-by-klaus-schwab.

Sindi, S. H., & Roe, M. (2017). *Strategic supply chain management.* Springer.

Stearns, P. N. (2018). *The Industrial Revolution in world history* (4th ed.). New York.

Stock, T., & Seliger, G. (2016). Opportunities of sustainable manufacturing in Industry 4.0. *Procedia Cirp, 40*, 536–541.

Tassel, L. (2019). *Why strive for Industry 4.0.* World Economic Forum Report (online). Available at https://www.weforum.org/agenda/2019/01/why-companies-should-strive-for-industry-4-0/.

Velasquez, N., Estevez, E., & Pesado, P. (2018). Cloud computing, Big data and the Industry 4.0 reference architectures. *Journal of Computer Science and Technology, 18*(3), e29–e29.

Venkatesh, V., Thong, J. Y., & Xu, X. (2012). Consumer acceptance and use of information technology: Extending the unified theory of acceptance and use of technology. *MIS Quarterly, 36*(1), 157–178.

Wilson, H.J., & Daugherty, P.R. (2018, August). Why even AI-powered factories will have jobs for humans. *Harvard Business Review* (online). Available at https://hbr.org/2018/08/why-even-ai-powered-factories-will-have-jobs-for-humans.

Big Data Analytics and Accounting Education: A Systematic Literature Review

Tamanna Dalwai, Syeeda Shafiya Mohammadi, Gaitri Chugh, and Alan Somerville

1 Introduction

Big data is referred to, as the outcome of the digital revolution (Moffitt, 2013). In recent times it has been described as "the new big thing". Crude has found a new synonym and that is none other than big data. The importance of big data is summarized by its ability to drive decision-making for corporate institutions, government agencies, nonprofit organizations, and even individuals. Its role as an enabler in resolving the bottlenecks across various business processes is researched and documented (Moffitt, 2013; Vasarhelyi et al., 2015). Business Intelligence (BI)

T. Dalwai (✉) · S. S. Mohammadi · G. Chugh
Muscat College, Muscat, Oman
e-mail: tamanna@muscatcollege.edu.om

S. S. Mohammadi
e-mail: syeeda@muscatcollege.edu.om

G. Chugh
e-mail: gaitri@muscatcollege.edu.om

A. Somerville
University of Stirling, Stirling, UK
e-mail: s.a.somerville@stir.ac.uk

© The Author(s), under exclusive license to Springer Nature
Singapore Pte Ltd. 2021
N. R. Al Mawali et al. (eds.), *Fourth Industrial Revolution and Business Dynamics*, https://doi.org/10.1007/978-981-16-3250-1_8

159

provides a paradigm shift in terms of providing innovative solutions and unlocking or creating value propositions, which are considered to be milestones in development of accounting education (Bhimani, 2014; Griffin, 2015).

Big data can be defined as giving meaning to the mess. A more refined definition of Big Data is its ability to manage the three V's—Volume, Variety, and Velocity (Gartner Inc., 2019). This processing of the three V's, enabling effective decision-making and leading to process automation coupled with creative insights has helped many stakeholders to make sense of both the quantity and quality of information available to analyze. Further research shows that components like accuracy and reliability of data referred to as Veracity along with Value creation are the additional two V's linked to Big Data (Merritt-Holmes, 2016). Historically, accounting was meant to facilitate decision-making (Capriotti, 2014) before the introduction of a double-entry book-keeping system in accounting by Luca Pacioli in the fifteenth century (Payne, 2013). The accountants are entrusted with the task of summarizing information in paper-based documents into meaningful and simple language for the decision-makers to take effective decisions.

Big Data is seen as the big thing of today, and it is believed that it is going to drive and define the view of tomorrow. The computer and data storage advancements, and the ability to collect such large quantities of unique and specific data, various advanced forecasting techniques can be applied in an almost infinite number of ways. With the requirements of various corporate organizations and government bodies, to predict the future trends with high accuracy, taking into account both high volume and high variety, this has taken Big Data to altogether new level (Merritt-Holmes, 2016). Big Data, and the follow-on concept of Business Analytics, have changed the way organizations work, as it is being used to help in providing answers to some key business questions such as the descriptive—"what happened?"; the diagnostic—"why did it happen?"; the predictive—"what will happen?"; and finally—to enable decision-making—the prescriptive—"what should be done?" (Gartner Inc., 2019).

Realizing the importance of Big Data and the value it adds, various professional accountancy associations and firms have strongly suggested the need for realignment toward a data-centric ecosystem. The need to develop an integrated framework is stronger than ever before, and various stakeholders are finding ways to integrate the Big Data with the accounting curriculum. Leading accounting and consultancy firms (such as PWC, 2015) have suggested frameworks on integrating Big Data, in order to improve the quality of financial audits, add value to the consultation process, and improve skills in areas such as risk management. There is enough evidence to support the viewpoint of integrating this level of information technology within the accounting curriculum framework (Behn et al., 2012; Lawson et al., 2013). Various researchers have recommended the development of a new Accounting Information system [AIS], to both upgrade study in accounting majors, and to integrate aspects of AIS in other aspects of accounting curriculum, so as facilitate "holistic learning" (Apostolou et al., 2014). Inclusion of such technology-highlighting aspects would aid the accounting firms in recommending and enabling the successful implementation of business strategies (Lawson et al., 2013).

As the role of accountants has become more focused on enabling support for decision-making, as opposed to processing of transactions in the past, it has become critical to integrate interactive data visualization in the financial accounting curriculum. As mentioned above, this would improve the engagement of students, making course work both more interesting and relatable (Janvrin et al., 2014). Practical and effective ways have been recommended for educators to integrate technology in teaching financial accounting, such as using different types of software based on spreadsheets (Boyer & Lyons, 2011). For areas such as management and cost accounting studies, wherein the data size is large and often both structured and non-structured, the use of different business intelligence software is required. Power BI, Click, and Tableau, where the data is depicted graphically and features interactive data analysis, it has been shown it can transform the learning process (Janvrin et al., 2014). The use of pivot tables, and the inclusion of data-sorting functions such as "add INS" (for example in a pivot table where data set from various sources can be compiled) is a very effective way to provide information. All of these functions are actively used by senior management to objectively and effectively drive the decision-making process (Bradbard et al., 2014). Tools like Tableau and Microsoft Access have also been found to aid in analyzing profits, and students can be trained to use these and other systems, to suggest recommendations based on the findings using specific profitability management software (Igou & Coe, 2016).

Information technology also plays a key role in the auditing process; the myth that business intelligence and data analytics cannot complement the audit process has been dispelled. With the advent of computer aided audit techniques, students are able to finalize an audit assignment (Daigle et al., 2011). Data visualization, using various business intelligence tools, is an asset for an auditor in analyzing an entity's business processes. One of the major software developments in the field of auditing, is Caseware, developed in collaboration with the IDEA academic partnership. It should be noted that different auditing firms have their own proprietary auditing software (Li & Chang, 2013).

Corporate reporting has undergone a drastic change in terms of the presentation and processing of the information, thanks to Big Data and business analytics, coupled with other modern technologies. It has become essential to incorporate the use of IT in the field of corporate reporting, as it empowers the accountant/analyst to enhance the level of reporting; the data interpretational quality is improved, owing in part to the various business intelligence and visualization tools. Certainly, with the general accounting curriculum focusing on specific skills, there is a need to integrate teaching with both modern technology and Big Data (Al-Htaybat & von Alberti-Alhtaybat, 2017).

Forensic accounting is now at the epicenter of modern world business transactions, and it has shown itself to be the main accounting service, to both detect fraud, and ensure the various regulations pertaining to corruption and bribery are in place (Crumbley & Stevenson, 2015). Demand for forensic accounting is increasing owing to various fraudulent situations becoming more apparent; for example, it was reported that each year companies lose about 5 percent of their revenues to fraud, this worldwide cumulatively exceeds US$3.5tn (Association of Certified Fraud Examiners, 2016). Forensic accountants are now progressively relying on Big Data (analytics) in practice to deal with company data sets, which help mitigate the constraints related to traditional spreadsheet (EY, 2016).

The above discussion highlights the importance of big data analytics in the accounting profession. However, there is still limited research across the two fields for developing current accounting education. To the best of the researcher's knowledge, there are no review studies that have explored the depth of big data analytics in the accounting curriculum. The primary goal of this chapter is to investigate the emerging themes from the academic research papers using a systematic literature review on big data analytics and accounting education. To achieve this aim, the review is structured as follows: Sect. 2 illustrates the research methodology adopted for conducting the systematic literature review; Sect. 3 presents the major findings of the review emerging for the selected academic journal articles; finally, Sect. 4 draws the conclusions, discusses the recommendations and limitations of this review.

2 RESEARCH METHODOLOGY

The research methodology consists of four distinct phases as shown in Fig. 1. The first step is establishing a review protocol. Secondly, for the related publications, the inclusion and exclusion criteria are defined. Third, an in-depth study search is conducted, followed by critical evaluation, data extraction, and synthesizing past findings.

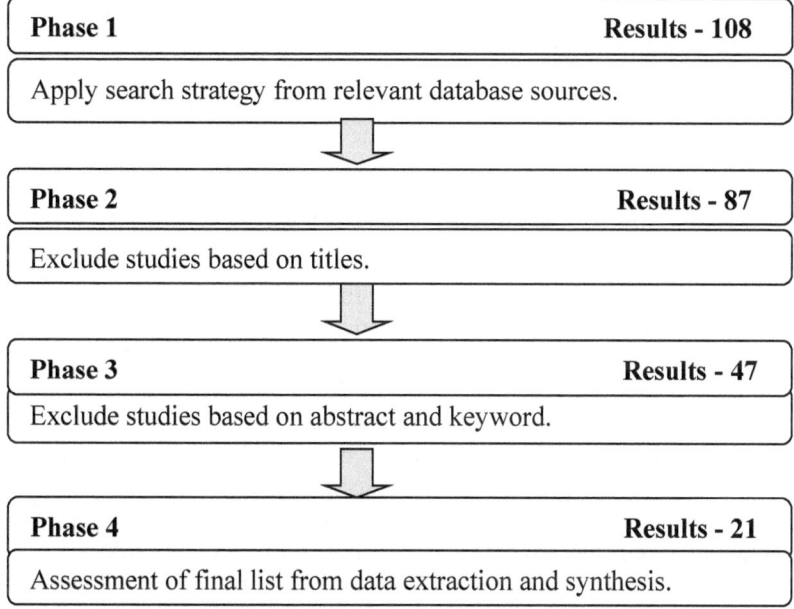

Fig. 1 The four phases of the systematic literature review (*Source* Created by authors using Kitchenham et al. [2009] systematic literature review)

2.1   Protocol Development

At the initial phase in the systematic review of literature, a protocol for successive phases was designed and developed. The study is based on the principles highlighted in Cochrane Handbook and Policies for Systematic Reviews of Intervention (Higgins & Green, 2011). The key objectives of this review, which describes and synthesizes past results in big data analytics and its application to accounting education, is discussed in this review protocol. Based on this objective, the criteria are created to take into account both the inclusion or exclusion of papers, as well as the search method, quality evaluation and the categorization of the results. By adopting a systematic literature review methodology, the various themes emerging from research articles on big data analytics and accounting curriculum are examined. The review process was thus driven by the research question: *How is the knowledge related to big data embedded in the accounting curriculum and what are the drivers supporting this initiative?* Thus, using these criteria, a number of parameters are therefore established in the study to categorize the papers (Mikalef et al., 2018).

2.2   Inclusion and Exclusion Criteria

The credibility of a systematic literature review is established by applying multiple criteria formats, for inclusion and exclusion of research articles in the selection phase. The research articles were chosen because of the focus on how big data could provide value in accounting education. The publications are selected from 2010 to 2019 as words big data gained momentum in the educational and business communities during this period (Mikalef et al., 2018). The systematic analysis covered academic papers, such as journal articles and conference papers.

The research utilized Scopus and Web of Science (WoS) as the most integrated electronic databases that support the identification of relevant Scopus indexed research articles. The Citation Index, Book Citation Index and Emerging Sources Citation Index of conferences are excluded (Maroufkhani et al., 2019). Finally, as the focus is on the incorporation of big data analytics in accounting education, empirical, qualitative and case studies with performance results were included.

2.3 Data Sources and Search Strategy

The data search was initiated from 10–10–2019 to 20–10–2019 and the search strategy began by generating search strings that were transformed into keywords. The search strings were used in two sets of keywords and the first being used as "big data" and other string sets being used as accounting, accounting education, accounting curriculum, data analytics, and graduate skills. The name, abstract, and keyword portions of the manuscripts were checked for keywords. The string words were searched in the electronic databases of Scopus and Web of Science. Later, these search keywords were uploaded in NVIVO for analysis.

Papers, articles and their specific rankings were reviewed. In this process, research that were not giving a clear perspective of Big Data Analytics in education were excluded, and only articles based on big data and accounting education were included.

2.4 Assessment of Final List from Data Extraction and Synthesis

In terms of the various quality criteria, each of the 47 papers left after phase 3 are independently evaluated by the authors. Webster and Watson (2002) have proposed that a high-quality evaluation focuses intensively on the principles discussed in the paper. Hence, each individual paper is coded according to the different criteria, ensuring credibility and validity of the papers being fully linked to the Big Data Analytics and accounting curriculum (Mikalef et al., 2018). At this point, 26 papers are not included, and 21 papers are left for data extraction and synthesis.

To synthesize the outcomes and identify studies based on their scope, an analysis is performed using the keyword "Big Data", in conjunction with results returning the following words—organizations, tangible and intangible resources, team management, the implementation and distribution of Big Data programs, governance of Big Data projects, and the ethics and morals of Big Data business issues.

The data extraction and publication categorization are developed during meetings of all researchers. In accordance with the coding system, the remaining 21 papers are extensively analyzed, and the relevant data collected, processed and synthesized.

The papers are manually coded by two researchers to solve discrepancies in coding. Coding is completed using NVIVO. For the purpose of coding, a text search query is run in NVIVO to run the manual codes.

Additionally, the recurring items in the selected papers are found using the word cloud in Fig. 2. A word cloud gives a visual representation to the most frequent words in the selected papers (Mariani et al., 2018). The words in Fig. 2 show a generic representation reflecting how subject area has been treated and suggests a relatively traditional, systematic and experimental approach.

Fig. 2 Word cloud with the frequently used terms in selected papers (*Source* Created by authors using NVIVO software)

3 RESULTS

3.1 Data Analytics Skills Required in Education

The first emerging theme arising from the systematic literature review was the data analytics skills. An investigation of the data analytics skills considered to be most important was collected through a survey administered to the accounting faculty (Dzuranin et al., 2018). The findings of this survey suggested that in the order of most to the least important data analytics skills were: (Microsoft) Excel or equivalent, analytical software packages for auditing such as IDEA and ACL, visualization packages such as Tableau, Qilk, Microsoft BI, database software, statistical programming packages, and other programming languages. Dzuranin et al. (2018) concluded that universities need to develop student's inquisitiveness that can be answered using data. This approach would result in enriching the accounting curriculum with analytical thinking and problem-solving skills.

A survey in 2015 by the Pathways Commission suggested that ERP was the third most important element for the student to learn, as this system automatically generates Big Data (Janvrin & Watson, 2017). Janvrin and Watson (2017) advised instructors to teach analysis of data by using different software packages, and also elevate student's understanding of the differences in Excel, visualization software and other analytic software. The authors also summarize the list of vendors that provide the free software for class use. For example, Big Data University offers free events and courses on different topics that cover R, Big Data, and Text Mining. Similarly, IDEA provides access to free generalized audit software, datasets, and a download book. A pro-analytics attitude has to be cultivated by schools and universities, in order to equip graduates with the skills required by organizations (Carillo et al., 2019). The results of Rezaee and Wang (2019) indicate the need for the integration of Big Data in forensic accounting education, and of developing advanced analytical data management skills, for extracting, transforming, and leveraging of data in forensic accounting practices.

3.2 Examples of Big Data Analytics Embedded in the Accounting Curriculum

Dzuranin et al. (2018) suggests three approaches for embedding the big data analytics in accounting curriculum: a focus approach, an integrated approach, and a hybrid approach. Through a focused approach, a stand-alone course is offered on the foundations of data analytical skills. The integrated approach proposes the infusion of an existing accounting course with data analytical skills. A hybrid approach takes a position of offering both a stand-alone course and an existing accounting course with embedded data analytics. PwC (2015) advocates introducing programming (such as Python or Java), structured and unstructured database management, and data visualization tools knowledge into courses. Rezaee and Wang's (2019) findings advocate the integration of Big Data in forensic accounting courses, at both the undergraduate and graduate levels. In financial accounting courses, students can be assigned tasks on researching company filings, listed by the Securities and Exchange Commission, which develops experiential learning (Sledgianowski et al., 2017). Overall, research argues for the integration of information technology at an analytical level, in introductory accounting courses, by using relevant spreadsheet software.

3.3   Professional Bodies and Accreditation Agencies Requirements on Data Analytics Skills

Dzuranin et al. (2018) discusses the requirements of the accounting programs having Association to Advance Collegiate Schools of Business (AACSB) accreditation that emphasizes the integration of data analytics in the curriculum. The AACSB (2016) requires colleges to think about integrating the accounting curriculum with statistics, data management, analytics and hands-on use of Big Data tools. Rezaee and Wang (2019) findings support the initiatives of the AACSB in integrating information technology throughout academic curricula.

The Pathways Commission was formed by the American Institution of Certified Public Accountants (AICPA) and the American Accounting Association (AAA), to examine the future of accounting education (Pathways Commission, 2012). The Commission recommended transforming accounting education with enumerated technologies such as data visualization, cloud infrastructures, integrated audit modules, and database dashboard metrics.

The Institute of Chartered Accountants In England and Wales (ICAEW) published a webcast on accounting and Big Data, and the speakers suggested that there should be caution on Big Data being considered as the future of accounting (Al-Htaybat & von Alberti-Alhtaybat, 2017). The expert panel members also voiced the fear that too much disclosure can affect the competitive advantages of the firm.

Gepp et al. (2018) however, advocate the need for harnessing big data techniques, and the methods for forecasting financial distress, which can improve the going concern evaluation in an audit (as required by the AICPA issued Statement on Auditing Standards, No. 59). The AAA annual meeting also discussed the need to update the accounting curriculum with data analysis courses (PwC, 2015).

Students are expected to be competent in both audit and tax advisory skills and should therefore be exposed to studies in how to be a data scientist. However, the current auditing standards do not take into account data analytical approaches (Earley, 2015). The Institute of Management Accountants (IMA) updated its competency framework for management accountants and expects the development of technology and analytical skills in their study skills (IMA, 2018).

3.4   Recommendations on Big Data Analytics Skills for Accounting Graduates

The findings of Dzuranin et al. (2018) suggest that data analytical skills should be embedded in the curriculum, using a hybrid approach. It would be useful for students to be taught data analytics skills using hands-on project and case studies. Mesa (2019) advocates investigating insights into teaching data analytics to accounting students by conducting research on students sense-making properties. The author also proposes examining how accounting assignments can foster iterative cycles. Al-Htaybat and von Alberti-Alhtaybat (2017) also recommend that future research is conducted using an empirical analysis of Big Data analytics use in accounting functions. Furthermore, the study suggests investigation of how data analytics, multimedia analytics, text analytics, web analytics and mobile analytics, can be used in corporate reporting. Others state that the accounting education agenda needs to address the technological developments, which will have practical insights (Al-Htaybat & von Alberti-Alhtaybat, 2017). Research on big data applications in auditing has suggested validating the effectiveness of big data uses in auditing (Gepp et al., 2018), given the premise that auditing can benefit from big data financial distress and financial fraud models.

Earley (2015) prefers caution while revamping the curricula. His research recommends that students are taught to comprehend and understand the relationship between the financial statement, business processes, and business risk. (Carillo et al., 2019) suggests preparing individuals for the data-driven era by the joint collaboration of businesses and educational institutions. Sledgianowski et al. (2017) recommend developing a fluid plan for accounting education, to adjust for competencies that accommodate topical priorities, or changing mission and resources of the institutions or departments.

4   CONCLUSION

The primary objective of this study evolving themes in the academic literature on big data analytics and accounting education. Using the Scopus and Web of Science databases, this chapter examined 21 articles using a systematic literature review. The findings were categorized on the basis of the data analytical skills needed to be taught, the professional and accreditation agencies' requirements on data analytics, examples of data analytics being embedded in the curriculum, and the recommendations that have emerged in these research articles. That the research papers in this area are limited, suggests a slow progress in how data analytics being embedded in the curriculum.

It is worth noting that whether it is financial accounting, management accounting, forensic accounting or auditing, data analytics has a scope for each area. A range of software or systems such as ERP, Excel, Tableau, R, SPSS, Stata, and others have been recommended by different research papers to be taught to the students. It is also strongly advocated that education should focus on the type of decision-making that can take place in accounting, using appropriately analyzed data. The professional bodies such as AICPA, AAA and IMA, have all advocated the integration of data analytical skills. Similarly, accreditation agencies such as AASCB strongly advocate the use of big data tools in the accounting qualification, so as to have graduates that meet the market requirements.

The recommendations of the research articles strongly suggest the building of analytical and decision-making skills modules for accounting graduates. A hybrid approach is suggested that incorporates big data analytics, using hands-on project and case studies materials. However, caution is being suggested while curricula undergoes revision, as students need to be able to understand the links between financial accounting, business risk and business processes. The over-riding message is to ensure that accounting students get the right exposure to data analytical skills, in order to make them competent, and give them the relevant employability skill sets.

This research has three main limitations. Firstly, only those research papers were selected that were available in two main literature databases—Scopus and Web of Science. These are comprehensive databases that cover a large number of journal and scientific research publications. However they are not considered historically complete, as repositories take longer to index. Secondly, research in the area of big data analytics and accounting are gaining momentum, thereby the publications that are being published while this chapter has been written are not being accommodated. Lastly, a systematic literature review cannot be considered as a panacea or the end of the road, but a beginning of the new pathways for the future.

Besides the above limitations, the findings of this chapter have implications for both higher education institutions and policymakers. Policy makers, such as accreditation agencies and professional bodies, have to consider the inclusion of data analytical skills, and its implication for practice. This would support in recommending the areas in which graduates have to develop skills. Higher education institutions have a duty to their students, to consider not only developing their usage of software skills, but also how to train them to interpret the data for decision-making.

References

AACSB. (2016). *Eligibility procedures and accreditation standards for accounting accreditation.* Retrieved from http://www.aacsb.edu/~/media/AACSB/Docs/Accreditation/Standards/2013-acct-standards-update.ashx.

Al-Htaybat, K., & von Alberti-Alhtaybat, L. (2017). Big Data and corporate reporting: Impacts and paradoxes. *Accounting, Auditing and Accountability Journal, 30*(4), 850–873. https://doi.org/10.1108/AAAJ-07-2015-2139.

Apostolou, B., Dorminey, J. W., Hassell, J. M., & Rebele, J. E. (2014). A summary and analysis of education research in accounting information systems (AIS). *Journal of Accounting Education, 32*(2), 99–112. https://doi.org/10.1016/j.jaccedu.2014.02.002.

Association of Certified Fraud Examiners. (2016). *Report to the Nations on Occupational Fraud and Abuse: 2016 Global Fraud Study.* Retrieved from Austin, USA. https://www.acfe.com/rttn2016/docs/2016-report-to-the-nations.pdf.

Behn, B. K., Ezzell, W. F., Murphy, L. A., Rayburn, J. D., Stith, M. T., & Strawser, J. R. (2012). The pathways commission on accounting higher education: Charting a national strategy for the next generation of accountants. *Issues in Accounting Education, 27*(3), 595–600.

Bhimani, A. a. W., L. P. (2014). Digitisation, 'Big Data' and the transformation of accounting information. *Accounting and Business Research, 44*(44), 469–490.

Boyer, B., & Lyons, B. (2011). Rivera Custom Cabinetry: Financial Statement Analysis Using Excel.

Bradbard, D. A., Alvis, C., & Morris, R. (2014). Spreadsheet usage by management accountants: An exploratory study. *Journal of Accounting Education, 32*(4), 24–30. https://doi.org/10.1016/j.jaccedu.2014.09.001.

Capriotti, R. J. (2014). Big Data: Bringing big changes to accounting. *Pennsylvania CPA Journal, 85*(2), 36–38.

Carillo, K. D. A., Galy, N., Guthrie, C., & Vanhems, A. (2019). How to turn managers into data-driven decision makers: Measuring attitudes towards business analytics. *Business Process Management Journal, 25*(3), 553–578. https://doi.org/10.1108/BPMJ-11-2017-0331.

Crumbley, L. L., & Stevenson, S. H. (2015). *Forensic and investigative accounting* (7th ed.). Commerce Clearing House.

Daigle, J. J., Daigle, R. J., & Lampe, J. C. (2011). Using ACL script to teach continuous auditing/monitoring: The Tremeg case. *Journal of Forensic and Investigative Accounting, 3*(2), 277–389.

Dzuranin, A. C., Jones, J. R., & Olvera, R. M. (2018). Infusing data analytics into the accounting curriculum: A framework and insights from faculty. *Journal of Accounting Education, 43*, 24–39. https://doi.org/10.1016/j.jaccedu.2018.03.004.

Earley, C. E. (2015). Data analytics in auditing: Opportunities and challenges. *Business Horizons, 58*(5), 493–500. https://doi.org/10.1016/j.bushor.2015.05.002.

EY. (2016). *Shifting into high gear: Mitigating risks and demonstrating returns: Global Forensic Data Analytics Survey 2016.* Retrieved from London, UK. https://www.ey.com/Publication/vwLUAssets/EY-shifting-into-high-gear-mitigating-risks-and-demonstrating-returns-63x82/$FILE/EY-shifting-into-high-gear-mitigating-risks-and-demonstrating-returns.pdf.

Gartner Inc. (2019). *Gartner glossary: Big Data.* Retrieved from https://www.gartner.com/en/information-technology/glossary/big-data.

Gepp, A., Linnenluecke, M. K., O'Neill, T. J., & Smith, T. (2018). Big data techniques in auditing research and practice: Current trends and future opportunities. *Journal of Accounting Literature, 40*, 102–115. https://doi.org/10.1016/j.acclit.2017.05.003.

Griffin, P. a. W., A. (2015). Commentaries on Big Data's importance for accounting and auditing. *Accounting Horizons, 29*(22), 377–379.

Higgins, J. P., & Green, S. (2011). *Cochrane handbook for systematic reviews of interventions* (Vol. 4). Wiley.

Igou, A., & Coe, M. (2016). Vistabeans coffee shop data analytics teaching case. *Journal of Accounting Education, 36,* 75–86. https://doi.org/10.1016/j.jac cedu.2016.05.004.

IMA. (2018). *IMA management accounting competency framework*. Retrieved from www.imanet.org/career-resources/managementaccounting-competenc ies?ssopc=1.

Janvrin, D. J., Raschke, R. L., & Dilla, W. N. (2014). Making sense of complex data using interactive data visualization. *Journal of Accounting Education, 32*(4), 31–48. https://doi.org/10.1016/j.jaccedu.2014.09.003.

Janvrin, D. J., & Weidenmier Watson, M. (2017). "Big Data": A new twist to accounting. *Journal of Accounting Education, 38,* 3–8. https://doi.org/10.1016/j.jaccedu.2016.12.009.

Kitchenham, B. A., Brereton, O. P., Budgen, D., Turner, M., Bailey, J., & Linkman, S. (2009). Systematic literature reviews in software engineering—A systematic literature review. *Information and Software Technology, 51*(1), 7–15.

Lawson, R. A., Blocher, E. J., Brewer, P. C., Cokins, G., Sorensen, J. E., Stout, D. E., Sundem, G. L., Wolcott, S., & Wouters, M. J. (2013). Focusing accounting curricula on students' long-run careers: Recommendations for an integrated competency-based framework for accounting education. *Issues in Accounting Education, 29*(2), 295–317.

Li, Y., & Chang, K.-C. (2013). Exploring the dimensions and effects of computer software similarities in computer skills transfer. In *Innovative strategies and approaches for end-user computing advancements* (pp. 99–118). IGI Global.

Mariani, M., Baggio, R., Fuchs, M., & Höepken, W. (2018). Business intelligence and big data in hospitality and tourism: A systematic literature review. *International Journal of Contemporary Hospitality Management, 30*(12), 3514–3554. https://doi.org/10.1108/IJCHM-07-2017-0461.

Maroufkhani, P., Wagner, R., Wan Ismail, W. K., Baroto, M. B., & Nourani, M. (2019). Big Data analytics and firm performance: A systematic review. *Information, 10*(7), 226.

Merritt-Holmes, M. (2016). *Big Data & analytics: The DNA to a successful implementation in 2016*. Retrieved from London, UK. https://dataanalytics. report/Resources/Whitepapers/40de07ff-696d-40cd-b05d-ae38f26429c3_ w2.pdf.

Mesa, W. B. (2019). Accounting students' learning processes in analytics: A sensemaking perspective. *Journal of Accounting Education, 48,* 50–68. https://doi.org/10.1016/j.jaccedu.2019.06.003.

Mikalef, P., Pappas, I. O., Krogstie, J., & Giannakos, M. (2018). Big Data analytics capabilities: A systematic literature review and research agenda. *Information Systems and e-Business Management, 16*(3), 547–578.

Moffitt, K. a. V., M. (2013). "AIS in an age of Big Data" 1. *Journal of Information Systems*, 1–19.

Pathways Commission. (2012). *Charting a national strategy for the next generation of accountants.* Retrieved from http://commons.aaahq.org/posts/a34 70e7ffa.

Payne, R. (2013). *From Pacioli to Big Data.* Retrieved from https://economia. icaew.com/opinion/june2013/from-pacioli-to-big-data.

PwC. (2015). Data driven: What students need to succeed in a rapidly changing business world. In PwC Firm Publication, New York, NY.

Rezaee, Z., & Wang, J. (2019). Relevance of Big Data to forensic accounting practice and education. *Managerial Auditing Journal, 34*(3), 268–288. https://doi.org/10.1108/MAJ-08-2017-1633.

Sledgianowski, D., Gomaa, M., & Tan, C. (2017). Toward integration of Big Data, technology and information systems competencies into the accounting curriculum. *Journal of Accounting Education, 38,* 81–93. https://doi.org/10.1016/j.jaccedu.2016.12.008.

Vasarhelyi, M., Kogan, A. a., & Tuttle, B. (2015). Big Data in accounting: An overview. *Accounting Horizons, 29*(2), 381–396.

Webster, J., & Watson, R. T. (2002). Analyzing the past to prepare for the future: Writing a literature review. *MIS Quarterly*, xiii–xxiii.

Electronic Business

Diffusion and Adoption of E-wallets in Oman for Sustainable Growth

Gopalakrishnan Chinnasamy, Preeti Shrivastava, and Nitha Siju

1 INTRODUCTION

1.1 Background of the Study

The combination of wireless telecommunication, multifunctional mobile gadgets, lifestyle and payment system has created new philosophy and possibility in the real-world transactions besides paying through cash and cards (Seetharaman et al., 2017). The payment system has changed over the period of time in many ways from coins and currencies to paper money, then to plastic cards and virtual money (Jones, 2014). E-wallet has emerged as a new method of money among the prevailing payment system. This new way of payment is easy, quick and more inclusive way to handle money and do financial transactions. It enables the people to connect across the country in digital commercial platform without

G. Chinnasamy · P. Shrivastava · N. Siju (✉)
Department of Business & Accounting, Muscat College, Muscat, Oman
e-mail: nitha@muscatcollege.edu.om

G. Chinnasamy
e-mail: gopalakrishnan@muscatcollege.edu.om

P. Shrivastava
e-mail: preeti@muscatcollege.edu.om

entering any shipment details, card numbers or even username and passwords (Paypal, 2019). E-wallet is known as digital wallet and it maintains the credentials of users for various websites and the payment mechanism encrypted with enhanced security (Rathore, 2016).

E-wallet is basically an electronic card that is used for online payments through gadgets such as computer or smartphone. It is required to be connected with the individual's bank accounts to make the payments like credit or debit card encrypted with high security. E-wallet is a type of prepaid account and it is stored with money for anticipated future online payments. This can be used for make payment for e-commerce website transactions, groceries, online purchases, ticket booking and others, without entering any banks account number or swiping the plastic currencies. E-wallet has software to store the individual's personal details like name, shipping details, payment type, amount to be paid, debit or credit card details and other required information for the online transactions but with intense security and encryption of the information (The Economic Times, 2019). The e-wallet account can be opened by the users by installing the software on their gadgets and enter the required personal data with user password. Once the account has been created, the account needs to be filled with any amount and retained till the transactions takes place in the online platform. The amount can be refilled through the bank account for the future transactions. Here the customer is not required to fill any forms on any websites as the current data's is already stored in the wallet database. The diffusion of technology-oriented payment methods addressing the needs and wants of consumers whose adoption of standard mobile payment system is still to be explored (Ezell, 2010).

1.2   Need for the Study

Geographically, the market growth can be found in the major regions of North America and Europe. There are countries which are encouraging and moving toward cashless transactions like Asia pacific, China, Japan and India. This high growth opportunity of using e-wallet has been increasing due to the usage of smartphone in this region (Elaina, 2019). PayTM highlights that its valuation has been increased by 4.7% in the last quarter 2019. This could be reason for moving toward cashless transactions and transactions without any charges. Nowadays, almost all the nations across the world are moving and encouraging toward cashless payment system by using online wallet payment to have the transaction which is transparent and to avoid any kind of black money activities/transactions. However, the current global e-wallet segment is anticipated to the compound average growth rate (CAGR) of 15% and expected to attain the market size more or less USD 2200 billion by end of 2018–2023 (Kenneth Research, 2019), which is promising.

Sultanate of Oman is pioneer in upholding the diversified stable economy from their oil-based economy in the gulf region and their economic evolution in the region. The Central Bank of Oman (CBO) apex body of the Oman financial system is committed to facilitate mobile based, hassle-free Mobile clearing system (Mpclear) through e-wallet in the Sultanate of Oman in cooperation with banking sector and other stakeholders. This will eventually encourage cashless economy at the national level aligned with National E-government strategy laid in the Vision 2020 as well in Vision 2040. The straight through processing (STP) ensures the intro-operational platform of e-wallet of banking sector in Oman. The Central Bank of Oman or CBO authorized the licensed entity to perform the responsibility according to the rules and regulations to facilitate the unified switching and routing services between payment service provider (PSP) as National Mobile Payment Solution (NMPS) in Sultanate of Oman (CBO, 2017). At present, almost all the banks in Oman are having this application and there are few players in e-wallet payment businesses are e-floos, bmwallet, NBO wallet, Oman pay, alizzwallet, etc.

1.3 Problem Statement

The success and the sustainability of current business and financial services are highly dependent on their payment facility (Kamboj, 2014; Kelkar, 2014; Mukherjee & Chakraborty, 2016). The popularity has arisen for e-payment especially e-wallet, because of its favorable and user-friendly characteristics than any other traditional payment system. It is convenient, flexible, easy to operate, secured, confidential, privacy, free of charges and has high efficiency (Singh & Singh, 2016). E-wallet system has gained the greater importance over a period and this has been diffused by e-commerce throughout the world. The efficient payment system is inevitable for the development of any individual as well as for highly competitive country (Zarmpou et al., 2012). The steps of bringing up e-payment system became national agenda is essential to improve the

productivity and contributes country's economy positively. The technologically developed countries like Japan, US and UK has fully developed
e-wallet payment system; whereas developing countries are also focusing
in this segment from the growth perspective (Yang et al., 2012). Nevertheless, the CBO has mentioned that Oman would attain high economic
development and diversity by intensively transferring from paper based
to e-payment system to increase the operational productivity and cost
efficiency (CBO, 2017). Oman has scored 63.61% on the 2018 Global
Competitiveness Report published by World Economic Forum (WEF)
from averaged 19.14 points from 2008. The Global Competitiveness
index 4.0, 2019 measures the national competitiveness covering 141
countries' economies and has ranked Sultanate of Oman as 7th among
Middle-East and North Africa and overall it has marked 53rd position
witnessing the growth trend of the country's economy (WEF, 2019). To
become number one among the neighbors, the sultanate needs to fulfill
the Vision 2040 by enhancing the payment system by adopting e-wallet
as the major way of doing transactions. However, various stakeholders in
Oman, especially the general public are still reluctant to adopt the use of
e-wallet payment mode (Al-Lawati & Fang, 2016). Hence, it is imperative to improve the level of willingness and confidence to adopt the
e-wallet and also needs to remove the challenges associated with technology services and by providing efficient services to the stakeholders. As
such, the current study is trying to examine the level of diffusion and
factors for adoption of E-wallet payment system which is prevailing in
Oman economy.

1.4   Research Questions

To attain the objectives of the study, the following research objectives
have been developed:

- What factors influence the consumers to adopt the e-wallet payment
  system?
- What are the challenges which prevent the diffusion of e-wallet
  payment system?
- What are the implications of the findings to the stakeholders of e-
  payment system?

1.5   Research Objectives

- The main objectives of the study are:
- To know the main factors that influence the consumers to adopt e-wallet payment system
- To identify the possible solutions to mitigate the challenges which prevents the diffusion of e-wallet payment system
- To find out the implications from the findings to the stakeholders of e-payment system for the diffusion and adoption of e-wallet payment system.

1.6   Scope of the Study

The adoption of technology innovation is constantly managed by businesses and consumers. As such, exploring the factors determines their intention and acceptance behavior, especially technologically growing nation like Sultanate of Oman where there is an economic growth and start of technology development like e-payment system. The current study focuses on end consumers in Sultanate. It is essential to study the issues pertaining to slow e-wallet adoption rate in Sultanate. The study covers various transactions by businesses and consumers using the technology-oriented e-payment system such as e-wallet.

2   Literature Review
and Analysis of Related Work

The technological innovation is the trump card for the businesses to edge over the market for their product or services in the highly competitive environment. However, the benefits of the technological innovations for the economic development can only be observed when these innovations extensively are diffused and adopted (Hall & Khan, 2002).

2.1   Diffusion of Innovation (DOI)

The diffusion process is that the innovation is communicated through various sources over the period among the public (Rogers, 1995). Rogers have proved this during 2003 by introducing diffusion of innovations (DOI) theory and this has been applied in various research fields in the management arena as well as other relevant research areas. The technology diffusion and adoption-based studies used DOI as their theoretical framework of their research (Turan et al., 2015). Further, Rogers (2003) explained that, the DOI takes into account the five innovation factors (relative advantage, compatibility, complexity, trialability and observability) that are perceived by people and predicts the level of acceptance and adoption of an innovation. These five attributes have a positive influence on the adoption except the complexity.

2.2   Technology Readiness Index (TRI)

Technology readiness is people's tendency to understand and use new technologies for attaining the goals in personal life and work life. The activities or process takes direction from the new technology and it could be applied in the market by changing the scientific research into applied research. The survival of the new technology relies on the level of acceptance and progression through the desired outcome (Wiese & Humbani, 2019). The payment system has been involved with lots of technology revolution and made the system easier with the support of service providers such as financial institutions. In evaluating the individual's technology readiness, most of the studies follow Parasuraman's scale (2000) of technology readiness index (TRI) model. TRI assesses readiness to accept new technology and follows the traits of optimism, innovativeness, discomfort and insecurity. The technology readiness is driven by the first two traits whereas the other two traits are its inhibitors. As per Parasuraman's (2000), TRI endorses that there is relationship between the intension to use the technology and people's readiness of technology. People prefer to use new technology where there is optimism with innovations and there is no discomfort level and insecurity exist (Walczuch et al., 2007).

According to Rose and Fogarty (2010), there is no direct relationship between readiness and the intention to accept new technology; however, the diffusion and adoption of new technology is based on the market acceptance of that technology. The payment system has undergone with lots of enhancement through innovative technologies such as NEFT, net banking, mobile payment, e-wallet and other payment system and they all have dominated the world payment system. It is considered that mobile payment services like e-wallets are the key economic driven technologies in the developing as well as developed nations (Humbani & Wiese, 2019).

Sultanate of Oman is one of largest economies in gulf region with regards to mobile users. Hence, it is important to adopt e-wallet payment system in order to develop the country's business in the competitive world.

2.3   E-wallet Adoption and Diffusion

The mobile device cannot be separated from daily life of common man and this has become foundation of any technological development. The mobile devices opened up an opportunity to the businesses to do innovation strategy to sustain in the market. The e-wallet is one among the mobile based technologically innovative payment method (Olsen et al., 2011). Mobile applications are one of the promising activities that use mobiles devices, technologies and internet connectivity. Sultanate is one among the growing nations in the usage telecommunication services and mobile technology (Ali Haider Mohammad Saifullah Sadi, 2010). The telecommunication sector in Sultanate is occupied by two major service providers namely Omantel and Ooredoo in the earlier stages and at present there are other service providers like Friendi, Renna and Awasr. All these companies are offering internet services with high quality to support the development of Sultanate by enhancing the financial system through e-payment side (Al-Lawati & Fang, 2016; Nair & Fasal, 2017).

The success of any new technological ideas like e-wallet are dependent on the acceptance of the users of the new ideas. The behavior toward the use of new technology like e-wallet is influenced mainly by perceived usefulness and perceived ease of use (Venkatesh & Davis, 2000; Wijayanthi, 2019). The technology acceptance model (TAM) and technology readiness index (TRI) are the concepts which offers a theory for fundamental understanding that users' attitude is important in accepting and diffusing new innovation (Fishbein & Ajzen, 2011; Rose & Fogarty, 2006). The consumers perception is the influencing factor in identifying the diffusion rate (DOI) of e-wallet instrument (Tan & Chen, 2008; Wijayanthi, 2019). However, there are no studies related to e-wallet readiness specifically for Oman.

Many literatures analyzed the impact of factors on the adoption of new technology services during the initial period (Ajzen, 2011; Liébana-Cabanillas et al., 2017; Mallat, 2007; Munoz-Leiva et al., 2017; Pham & Ho, 2015; Rose & Fogarty, 2006; Wijayanthi, 2019). Moreover, majority of these studies related to adoption of mobile payment applications related to developed economies, whereas there are only few studies that address the status of developing countries mobile payment applications adoption. Similarly, the studies related to continued use of technology like mobile payment applications also has been discussed more on the developed nation perspective than developing and under-developed economies (Greenland & Kwansah-Aidoo, 2012; Wiese & Humbani, 2019; Wu et al., 2017). The outcome of these studies contradicts each other in terms of the adoption of mobile e-wallet apps among these economies. These conflicting outcomes needs to be studied further in the context of Oman and understand the factors which are influencing its adoption rate or factors which are inhibiting the acceptance and adoption of mobile e-wallet applications.

As per Lee (2009), Low and Harvey (2011), Venkatesh and Bala (2008), Wiese & Humbani (2019) TRI and DOI are the models that have been used by most of the researchers as theoretical framework in analyzing their research objectives. However, both of these methods are not utilized together to assess the level of adoption and diffusion technology innovation in the field of modern e-wallet mobile payment system. Hence, the present study is combining these models into one to assess the diffusion as well as adoption level of E-wallets in Oman. The following Fig. 1 depicts the theoretical framework for this study.

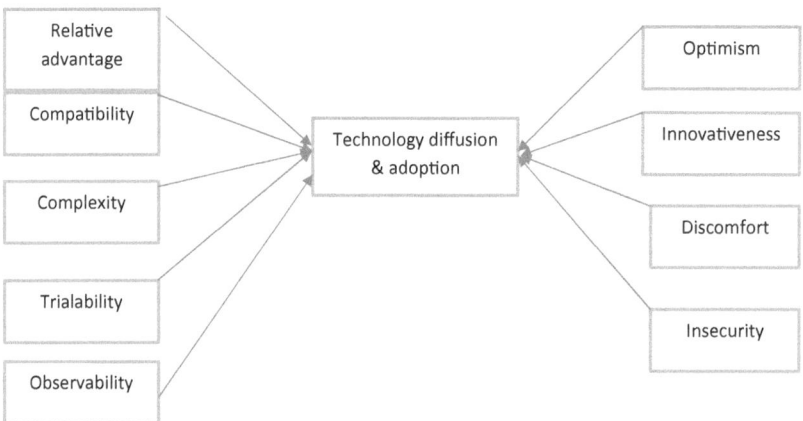

Fig. 1 Diffusion and adoption model (*Source* Theoretical model prepared by the authors)

3 Research Methodology

As mobile technology particularly e-wallet is relatively new research area in Sultanate of Oman with limited research contribution on the subject matter, thus the present study is using qualitative approach with focus groups interviews to explore the diffusion and adoption of e-wallet in Sultanate for the economic development of the nation (Al Lamki, 2018; Calder, 1977; Jarvenpaa & Lang, 2005). The respondents of this study are the users of e-wallet payment system and the service providers of this technology. The focus group interview has dynamism and team synergy gaining the momentum of the research area to bring out the effective outcome for the benefit of the society (Jarvenpaa & Lang, 2005). The optimal formations of groups are important to have more debatable discussions and interaction. According to Stewart et al. (2009) the degree to which the respondents feel comfortable in expressing the ideas and opinions determines the validity of focus group data. To keep in mind these points, the current study employs five groups consisted of different occupations and age categories like business group, service industry group, youngster group, middle aged group and service providers group (Gaizauskaite, 2012; Nyumba et al., 2018). The groups were again reformed later based on gender so that they can discuss freely. The total numbers of respondents were 30, which is a good number for focus group composition (Wilkinson, 2004). The researchers first did pilot study with one group and ensured the reliability of the interview questions and this group has not been included in the main study.

This study had some limitations in terms of finding the group discussion timings, availability of group members, the place of discussion, making minutes and so on. However, these have been resolved properly in the following manner, like in advance the date and time has been communicated earlier through emails and personal phone calls, and confirmations have been made. Those who were not able to participate in person due to their personal reasons have been provided with video chat room facility to participate in the group discussion. The group discussion has taken place in the common place during the weekends and it took place for about three hours. As the participant were not interested in allowing voice/video recording facility and as well as to maintain confidentially of the persons involved in the data development process the discussion points have been noted by the researchers and it has been maintained with confidentiality. Personal names and any identification related to job was not done while conducting discussion forum. Two researchers acted as moderators for the group discussions. The language used throughout the discussion was English.

4   RESULTS

The focus group discussions uncovered the outcome related to the e-wallet adoption among the people of Oman. The discussions highlighted that the people are aware of mobile phone services and also the expectations related to the services. The devices are majorly used for calls and social media usage purposes. Whereas the SMS and wireless application protocol (WAP) services were not used regularly by the group. Further, majority of the peoples in the group have experience in using e-wallet for their purchases. It is observed in the discussions that this purchase includes travel tickets, purchase from grocery shops/malls, utility bills and chain stores like Starbucks, Costa's, R&B etc. Interesting outcome was that purchases were made occasionally on a trial basis but not as a regular mode of payment. With this prevalent situation, the following are the discussion and outcome of the e-wallet payment characteristics, which has emerging factors of diffusion and adoption of e-wallet in Oman.

4.1   Relative Advantages

The relative advantages of e-wallet payments are plenty according to the groups opinion shared during the discussion forum. It includes the way to do purchases, place to do purchases, time to do purchase and avoid being in queues and crowds. They perceived that it is highly useful to do the transactions remotely without going to the place of sales. They feel it is more convenient it terms of safety, easy, quick and fair responses (IFC, 2018). Few discussion points are mentioned below which the group reflected:

> "One small device with a friendly application helps a lot and makes everything easier and possible in a moment. What more you want it? Just think about using cash or cards instead of using mobile wallet. Oops it's tedious for me when I am busy or in urgency. It is just simple and easy to have my e-wallet than take my wallet out and find a cash or card".

"Some time we used to forget to bring our wallet, but we never forget our mobile phone; it is obvious to use this as mode of payment".

"Most of the time we do not go with enough cash or coin and may not have time to take cash from ATMs or it may not be available at the access point and it may potentially create problems in our purchase mood. Hence, it is essential to have any one form of payment more specifically e-wallet which is not time consuming and is easy and friendly to use".

Hence, it is observed that group felt that e-wallet has more advantageous in terms of time, convenient, flexible, easy and helpful during busy time (He & Mykytyn, 2007). Moreover, more specifically it has increased the people's level of trust toward e-wallet payment system. This outcome is also supported by earlier researcher like Parasuraman (2000) and Rogers (2003) studies.

4.2   Compatibility

This particular component was evaluated in terms of how this payment system is well suited with different type of purchases at different situations. The group discussions main points are as follows:

"I would keep the payment through e-wallet for those expensive transactions and not keep those for buying chocolate, milk or eggs".

"I believe that e-wallet has convenience in a difficult situation like distance-wise, queues or where you don't have selling point".

"Imagine that in taking ticket for cinema, or for a shake you have to leave flat, take a car and be in queue and have the ticket. It is high pressure when time is limited. Better let's do with e wallet payment and utilize time".

"Think about the moment it would be more sensible to buy Air tickets, paying fees or vehicle fine, utility bills or any kind of utilities wherever it is possible to use".

The outcome of the discussions witnessed that there is a mixed response from the people who is willing to use the e-wallets. There is difference in the type of purchases either it is high valued purchases or routine kind of small buying activities. It is observed that they never mind about the types of purchases and they look for only comfortable way of shopping experiences either it can be 10 Omani Rial or 100 Omani Rial irrespective of the amount that is spend for the purchases. Interestingly, the discussion went like at what extent it has been used in a week or a month and it has been observed that one by third (approximately 33%) of their weekly or monthly purchases has been made through mobile payment and by using e-wallet (Humbani & Wiese, 2019).

4.3   Complexity & Discomfort

The group had discussions regarding the complexity and discomfort in adopting the e-wallets in their mobile phones. The discussion mentioned that complexity of e-wallet payment applications and mobile phones:

"I feel discomfort in procedure followed for the registration and payment arrangements. It seems complex. Sometime, the separate account system, separate billing arrangement creates more burden and feel it is highly irritating process".

"The language issues which I am facing some time while making payment. Though I am good in English language to read, write, speak, I need applications in my own language Arabic with more clear statements or translating options".

"There is much difference in the navigation key, various payment codes, payment service numbers and operational keys are different when compared to one to other applications".

"The fund transfer process is also annoying in terms of sharing the bank details and register to start the process are find it complex and creates more discomfort to me personally".

"It was difficult process. I did not find the clear instructions on how to do it one place. Sometime, I feel too much information without any clearer directions".

"I don't know anything about the applications, downloading, installing and how to use it easily. But I am feeling that it is not too late to learn and adapt it".

The above discussions are associating with the findings of the study related to diffusion and adaption of innovative technology models of Parasuraman (2000) and Rogers (2003).

4.4   Trialability & Observability

As stated by Rogers (2003), trialability and Observability is the opportunity to experiment the new technology introduced and make clear visibility in the market for a particular time period to its actual adoption or rejection. Further, the diffusion of technology is important part in encouraging the adoption. Hence, it is essential for the company who is introducing the new technology like e-wallet regarding the acceptance or rejection of the innovative ideas with proper reasons. The clear visibility regarding the new e-wallet applications also has been discussed in the forum. These two statements have been raised by the session moderators on discussion board to get more insight. Surprisingly, this has not been discussed much and there was no interest shown by these groups to escalate this discussion further. Still there are some comments as mentioned below:

"Even I am not aware how I learned it and not aware how I have started it. It is not a big deal to see whether this is convenient or not. I can switch anytime without any problem if I observe any difficulty in using it".

"I don't know how quick I became adapted to doing it. I just pay for the things even minding about it anymore I am so happy and comfortable when I use my mobile as wallet to pay my utilities in a cool manner. I think I am more inclined towards it nowadays".

Hence, it is observed that trailability and observability was not a significant predictor in adopting new technology and the intention to use it. The same trend has been observed in e-wallet payment technology adoption and diffusion (Nair & Fasal, 2017). Further, it has been inferred that trailability and observability encourages the consumer readiness and it has a stronger effect at the initial stage, but it's negatively affect the behavior of adoption for new technology at the later stages. It affects the diffusion of new technology positively (Al-Lawati & Fang, 2016; Nair & Fasal, 2017; Wijayanthi, 2019).

4.5   Optimistic and Innovative Customers

Surrounded with their self-managing ability in using new technology, optimistic and innovative customers will adopt the revolution in their routine payment system. These customers take action or moves toward new innovative ideas like e-wallet. A small optimistic approach can encourage them to migrate to new payment system very easily. This can be achieved by producing user friendly, easy adopting with no complex approach in adopting e-wallet. The discussion was taken place in the forum and there were some interesting points observed.

"It is easy to do the payment and get everything is possible without spending much time and boring to remember lots of credentials towards my payment of small amount".

"I may use these payment option later because it could be more famous and popular and most of us will be moving towards that. Moreover, my friends, relatives will start using when it becomes mandatory".

"Few seconds ago I got a message stating that the balance available in my e-wallet made me aware about my current financial state".

"I like that kind of technology or application or system has to be finished the activity before my screen goes to sleep mode. This kind of application I am looking for my payment activities".

The above statements explain few facts related to adopting e-wallet payment system. It is clearly observed that people are migrating toward new system sometimes without conscious effort. These may be their forward and confidence mindset and the tendency to accept innovative ideas. It is attractive and useful one for their social life (Al-Lawati & Fang, 2016; Wijayanthi, 2019). As such, we can infer that optimism and innovativeness are the influencing factors toward the diffusion and adoption of e-wallet payment system.

4.6   Insecurity

Insecurity may be defined as individuals with low level of confidence in new technology security features and a kind of pessimistic opinion about trying any new technologies like e-wallet. There are studies which indicate that there is a significant relationship between insecurity and e-wallet adoption until if they able to clearly understand the security feature. These kinds of perceptions are majorly occurring among middle aged people when compared to young people. This was observed in the focus group discussions also.

"Still, I am worrying and I am not convinced that the online threats and insecurity in using these kinds of payment system are not there. Please imagine if you lose your phone, how you will manage the situations".

"How do I select the secured e-wallet application, whereas this has been offered by any organization willing to offer the services to so many peoples? Somehow, I cannot convince myself if the services offered by any banks in Oman are totally safe".

"I would choose if the brands of the service providers are popular and familiar, since it has no connectivity with any popular banks".

"What extent the service providers are trustworthy to share all my credentials with them, like sharing the secrets".

"Sometimes I may make mistakes in filling the required credentials and this may lead to lose money and it is not at all possible to revert back. I feel it is a big challenge".

The new technology like e-wallet application could be extraordinary practices specially to middle age peoples to start practicing it because they have may have no experiences or very little experiences (Al-Lawati & Fang, 2016; Nair & Fasal, 2017; Wijayanthi, 2019). The other reason is that due to some bad experiences such as fraudulent transactions while making their online payment through online banking or card payment options. All these experience by individually or hearing from other experiences made them fear about the use of e-wallet application (Al-Lawati & Fang, 2016; Nair & Fasal, 2017; Wijayanthi, 2019). This should be handled well by making awareness to change their misperceptions about e-wallet usages. Hence, we can infer that insecurity factors negatively influences the diffusion and adoption of e-wallet in Oman.

5   CONCLUSION

The main objective of this paper was to identify the factors that affect the diffusion strategy used and the level of adoption among the users of e-wallet payment system in Oman. There are some positive adoptions of e-wallet payments related to different benefit areas which the services can be diffused to make the economy into digital economy (OECD, 2019). The advantageous aspects of e-wallet are that it is independent of time and place, remote and easy access to e-wallet payment ability. Further, this will bring the shopping experience more comfortable and no complex by avoid the queues and cash or card payment challenges (OECD, 2019). The findings according to the outcome from the focused group represent that the factors which are influencing the diffusion and adoption of e-wallet in certain environment where we need to manage the unexpected requirement of payment, helps in time savings maintaining the coins and currencies. The most effective payments vary according to the outcome from the focus group discussion is that the application can be used according to the convenience either it is 10 rial Omani or 100 rial Omani irrespective of the amount spending for their purchases. Interestingly, it has been observed that this application have been used by the respondents in a week or a month and it has been observed that one by third (approximately 33%) of their weekly or monthly purchases has been made through mobile payment using e-wallet (Wiese & Humbani, 2019). This is the positive step toward adoption and diffusion of e-wallet among Oman community.

There are factors preventing e-wallet payment adoption due to some specific influencing factors which has been derived from the focus group discussion such as security, complex application practices, not comfortable to use the application and it is not user-friendly method of payment. These findings will help the stakeholders to channelize their efforts to solve these issues to promote the use of e-wallet all over the nation by integrating the existing financial structure with telecommunication services. The emerging market is expecting the compatible application services by providing common standardized services from all the service providers to feel more convenient to use it. The awareness is also highly required to build the confidence toward trustworthy branded service providers, to remove any security issues pertaining toward the use of this application (Wijayanthi, 2019). This can be achieved in many ways but the main attention should take place by the stakeholders to ensure that error free or fraud free transactions are available at all time. Further, the motivational factors like offers, discounts, additional services, added benefits, integrated models are required to spread the acceptance and awareness about e-wallet (Pal et al., 2019). To penetrate the application fully in Oman market and it is imperative to familiarize the e-wallet in government sectors in paying bills, dues and other related payments.

As the study is qualitative in nature and presented with focus group discussion in Omani context, the findings of this study cannot be generalized to all the areas of the payment system and it is not the voice of the country as whole (Caillaud & Flick, 2017). This study will give an insight about the level of diffusion and adoption of the digital payment system and the required attention to encourage the new application in the transactions and payment platform. Further, this study can be extended geographically and other research approaches like quantitative or mixed method of research can be done to get more fruitful outcome for the development of this new technology Sultanate of Oman.

REFERENCES

Ajzen, I. (2011). The theory of planned behaviour: Reactions and reflections. *Psychology & Health, 26,* 1113–1127. https://doi.org/10.1080/08870446.2011.613995.

Ali Haider Mohammad Saifullah Sadi. (2010). The prospects and user perceptions of m-banking in the Sultanate of Oman. *Journal of Internet Banking and Commerce, 15*(2), 1–11.

Al-Lawati, B. H., & Fang, X. (2016). Diffusion of innovations: The case study of Oman's e-payment gateway. In *International conference on HCI in business, government, and organizations* (pp. 483–490). Springer.

Al Lamki, Z. (2018). The influence of culture on the successful implementation of ICT projects in Omani E-government.

Caillaud, S., & Flick, U. (2017). Focus groups in triangulation contexts. In R. Barbour & D. Morgan (Eds.), *Advances in focus groups research* (pp. 155–177). Palgrave Macmillan.

Calder, B. J. (1977). Focus groups and the nature of qualitative marketing research. *Journal of Marketing Research, XIV,* 353–364.

CBO. (2017). *Mobile payment clearing and switching system operating rules (online).* Available: https://cbo.gov.om/Pages/Payment-Systems-Operating-Rules.aspx.

Elaina Ransford. (2019). Increasing Adoption of Mobile Payments: Here's How Digital Banks Can Help. Retrieved from https://www.helpshift.com/blog/increasing-adoption-mobile-payments/. Accessed on 21.12.2019.

Ezell. (2010). *Contactless mobile payments.* The Information Technology & Innovation Foundation.

Fenchi Melissa, Cheng, Chamroeun, Khim, Sivmey, Thai. (2018). Consumer Adoption of E-Wallets: A Study of Millennials at the Institute of Foreign Languages, Cambodia, Proceedings of the 21st Asia-Pacific Conference on Global Business, Economics, Finance & Social Sciences (AP18Taiwan Conference) Taipei-Taiwan. December 21–22, 2018. Paper ID: W812.

Fishbein, M., & Ajzen, I. (2011). *Predicting and changing behavior: The reasoned action approach.* Taylor & Francis.

Gaizauskaite, I. (2012). The use of the focus group method in social work research. Socialinis Darbas Social Work, Mykolas Romeris University, 11(1), 19–30, ISSN 1648–4789 (print), ISSN 2029–2775 (online).

Greenland, S., & Kwansah-Aidoo, K. (2012). The challenges of market research in emerging markets: A practitioner perspective from Sub-Saharan Africa. *Australasian Journal of Market & Social Research, 20*(2), 9–22.

Hall, B. H., & Khan, B. (2002). Adoption of new technology. *New Economy Handbook,* 38.

He, F., & Mykytyn, P. (2007). Decision factors for the adoption of an online payment system by customers. *International Journal of Economics and Business Research, 3*, 1–32.

Humbani, M., & Wiese, M. (2019). An integrated framework for the adoption and continuance intention to use mobile payment apps. *International Journal of Bank Marketing*.

International finance corporation (IFC) World Bank group. (2018). Digital access: The future of financial inclusion in Africa service user, in Lusaka, Zambia.

Jarvenpaa, S. L., & Lang, K. R. (2005). Managing the paradoxes of mobile technology. *Information Systems Management, 22*(4), 7–23.

Jones, G. (2014). What is money? From commodities to virtual currencies/bitcoin. *SSRN Electronic Journal*, 1–12.

Kamboj, S. (2014). Financial inclusion and growth of Indian economy: An empirical analysis. *The International Journal of Business & Management, 2*(9), 175–179.

Kelkar, V. (2014). Financial inclusion for inclusive growth. *ASCI Journal of Management, 39*(1), 55–68. Retrieved from http://journal.asci.org.in/Vol. 39(200910)/391Vijay%20Kelkar.pdf.

Kenneth Research. (2019). *E-wallet market research report—Global forecast to 2023*. Retrieved from https://www.kennethresearch.com/report-details/e-wallet-market-/10159146. Accessed on 21.12.2019.

KPMG. (2019). *Press release, Oman banking sector demonstrates healthy growth (online)*. Available: https://home.kpmg/om/en/home/media/press-releases/2019/05/oman-banking-perspectives.html.

Lee, M. C. (2009). Factors influencing the adoption of internet banking: An integration of TAM and TPB with perceived risk and perceived benefit. *Electronic Commerce Research and Applications, 8*, 130–141.

Liébana-Cabanillas, F., Ramos de Luna, I., & Montoro-Ríos, F. (2017). Intention to use new mobile payment systems: A comparative analysis of SMS and NFC payments. *Economic Research-Ekonomska Istraživanja, 30*(1), 892–910.

Ligon, E., Malick, B., Sheth, K., & Trachtman, C. (2019). What explains low adoption of digital payment technologies? Evidence from small-scale merchants in Jaipur, India. *PLoS ONE, 14*(7), e0219450. https://doi.org/10.1371/journal.pone.0219450.

Low, H. Y., & Harvey, N. (2011). Shopping without pain: Compulsive buying and the effects of credit card availability in Europe and the Far East. *Journal of Economic Psychology, 32*(1), 79–92.

Mallat, N. (2007). Exploring consumer adoption of mobile payments—A qualitative study. *The Journal of Strategic Information Systems, 16*(4), 413–432.

Mukherjee, A., & Chakraborty, S. (2016). Financial inclusion of the poor and marginalized in Jharkhand: Analysis of theexisting model. *International*

Journal on Research and Development: A Management Review, 1(1). Retrieved from http://papers.ssrn.com/sol3/papers.cfm?abstractid=2169673.

Munoz-Leiva, F., Climent-Climent, S., & Liébana-Cabanillas, F. (2017). Determinants of intention to use the mobile banking apps: An extension of the classic TAM model. *Spanish Journal of Marketing-ESIC, 21*(1), 25–38.

Nair, R. S., & Fasal, S. (2017). Mobile banking and its adopting challenges. *International Journal of Computer Applications, 160*(4).

Nyumba, T., Wilson, K., Derrick, C. J., & Mukherjee, N. (2018). The use of focus group discussion methodology: Insights from two decades of application in conservation. *Methods Ecol Evol, 9*(1), 20–32.

OECD. (2019). *Southeast Asia going digital: Connecting SMEs*. OECD.www.oecd.org/going-digital/southeast-asia-connecting-SMEs.pdf.

Olsen, M., Hedman, J., & Vatrapu, R. (2011, June). E-wallet properties. In *2011 10th International Conference on Mobile Business* (pp. 158–165). IEEE.

Pal, A., Herath, T., & Rao, H. R. (2019). A review of contextual factors affecting mobile payment adoption and use. *Journal of Banking and Financial Technology, 3*(1), 43–57.

Parasuraman, A. (2000). Technology Readiness Index (TRI): A multiple-item scale to measure readiness to embrace new technologies. *Journal of Service Research, 2*(4), 307–320.

Paypal. (2019). *Privacy statement of paypal account (online)*. Available: https://www.paypal.com/en/webapps/mpp/ua/privacy-full#2.

Pham, T. T. T., & Ho, J. C. (2015). The effects of product-related, personal-related factors and attractiveness of alternatives on consumer adoption of NFC-based mobile payments. *Technology in Society, 43*, 159–172.

Rathore, H. S. (2016). Adoption of digital wallet by consumers. *Bvimsr's Journal of Management Research, 8*(1), 69–75.

Rogers, E. M. (1995). Diffusion of innovations: Modifications of a model for telecommunications. In *Die diffusion von innovationen in der telekommunikation* (pp. 25–38). Springer.

Rogers, E. M. (2003). *Diffusion of innovations* (5th ed.). Free Press.

Rose, J., & Fogarty, G. J. (2006). Determinants of perceived usefulness and perceived ease of use in the technology acceptance model: Senior consumers' adoption of self-service banking technologies. In *Proceedings of the 2nd Biennial Conference of the Academy of World Business, Marketing and Management Development: Business Across Borders in the 21st Century* (Vol. 2, pp. 122–129). Academy of World Business, Marketing and Management Development.

Rose, J., & Fogarty, G. (2010). Technology readiness and segmentation profile of mature consumers. *Academy of World Business, Marketing & Management Development, 4*, 57–65.

Stewart, D. W., Shamdasani, P. N., & Rook, D. W. (2009). Group depth interviews: Focus group research. *The SAGE Handbook of Applied Social Research Methods, 2,* 589–616.

Singh, D., & Singh, H. (2016). Market penetration by Indian Banks: Motives and motivators. *Indian Journal of Financ, 10*(3), 28–42.

Seetharaman, A., Kumar, K. N., Palaniappan, S., & Weber, G. (2017). Factors influencing behavioural intention to use the mobile wallet in Singapore. *Journal of Applied Economics & Business Research, 7*(2).

Tan, W. K., & Chen, S. K. (2008). An analysis of the factors influencing success of bank-issued micropayment system in Taiwan. *Journal of Systems and Information Technology.*

The Economic Times. (2019). *Definition of 'E-wallets' (online).* Available: https://economictimes.indiatimes.com/definition/e-wallets.

Turan, A., Tunç, A. Ö., & Zehir, C. (2015). A Theoretical Model Proposal: Personal Innovativeness and User Involvement as Antecedents of Unified Theory of Acceptance and Use of Technology. *Procedia-Social and Behavioral Sciences, 210,* 43–51.

Venkatesh, V., & Bala, H. (2008). Technology acceptance model 3 and a research agenda on interventions. *Decision Sciences, 39*(2), 273.

Venkatesh, V., & Davis, F. D. (2000). A theoretical extension of the technology acceptance model: Four longitudinal field studies. *Management Science, 46*(2), 186–204.

Walczuch, R., Lemmink, J., & Streukens, S. (2007). The effect of service employees' technology readiness on technology acceptance. *Information & Management, 44*(2), 206–215.

Wiese, M., & Humbani, M. (2019). Exploring technology readiness for mobile payment app users. *The International Review of Retail, Distribution and Consumer Research, 30,* 1–20.

Wijayanthi. (2019). Behavioral intention of young consumers towards e-wallet adoption: An empirical study among Indonesian users. *Russian Journal of Agricultural and Socio-Economic Sciences, 85,* 79–93. https://doi.org/10.18551/rjoas.2019-01.09.

Wilkinson, S. (2004). Focus group research. In D. Silverman (Ed.), *Qualitative research: Theory, method and practice* (2nd ed.). Sage.

World economic forum (WEF). (2019). *The Global Competitiveness Index 4.0 2019 Rankings (online).* Available: http://www3.weforum.org/docs/WEF_GCR_2019_Rankings.pdf.

Wu, J., Lin, L., & Huang, L. (2017). Consumer acceptance of mobile payment across time: Antecedents and moderating role of diffusion stages. *Industrial Management & Data Systems, 117,* 1761–1776. https://doi.org/10.1108/IMDS-08-2016-0312.

Yang, S., Lu, Y., Gupta, S., Cao, Y., & Zhang, R. (2012). Mobile payment services adoption across time: An empirical study of the effects of behavioral beliefs, social influences, and personal traits. *Computers in Human Behavior, 28*(1), 129–142.

Zarmpou, T., Saprikis, V., Markos, A., & Vlachopoulou, M. (2012). Modeling users' acceptance of mobile services. *Electronic Commerce Research, 12*(2), 225–248.

Present Practices and Future Challenges in Social Media Usage for Business: Observations from the United Arab Emirates

Ashavaree Das and Shreesha Mairaru

1 Introduction

In the era of the internet, Social Media (SM) is emerging as a powerful tool for users to create, exchange, and share content. A number of SM platforms are available today, each with its own distinctive features. SM platforms are based on interactive Web 2.0 based technologies that help create community by enabling users to consume and generate content simultaneously.

SM users have increased exponentially in the last few years. According to a study conducted by 'Hootsuite' and 'We Are Social' websites, 4.021 billion people around the world are using the internet in 2018. The research indicates that the number of internet users has gone up by 7% when compared to 2017. However, in 2018 the number of SM users has increased by 13% to 3.196 Billion. According to 'Stastica' and 'The

A. Das · S. Mairaru (✉)
Applied Media Division, Dubai Women's College, Higher
Colleges of Technology, Dubai, United Arab Emirates
e-mail: smairaru@hct.ac.ae

A. Das
e-mail: adas1@hct.ac.ae

© The Author(s), under exclusive license to Springer Nature
Singapore Pte Ltd. 2021
N. R. Al Mawali et al. (eds.), *Fourth Industrial Revolution and Business
Dynamics*, https://doi.org/10.1007/978-981-16-3250-1_10

Next Web', as of 2018, Facebook has 2.26 billion account holders, and YouTube has 1.9 billion subscribers. Facebook and YouTube hold the top two positions among SM platforms. If these SM platforms are considered as a separate country, then Facebook and YouTube are the most populated countries in the world today. In the UAE, almost 99% of the total population are active on SM. The annual growth in active SM users is a staggering 6.3% and the average time spent on SM is almost 3 hours (UAE SM Usage Statistics, 2020). With such a large number of people, SM has gained the potential to influence every walk of human life. Business is no exception. For brands heading into 2020 and beyond, there is no better time to reflect and refocus on how they approach SM. To that end, the authors highlight how companies have made sense of their social data and note current practices in five areas—marketing; reputation management; customer management; crowdsourcing; and employee relations. Furthermore, they offer an overview of challenges and opportunities for brands to approach SM moving forward into a new decade. Additionally, these practices are situated within the UAE context by exploring how small and medium firms use SM in their business agenda.

1.1   Objectives of the Study

This study aims to:

- Understand the theoretical perspectives that elucidate current SM usage trends and future challenges in business and marketing.
- Compare the global SM usage trends in the field of business with the United Arab Emirates.
- Outline best practices in SM and challenges for businesses in 2020 and beyond.

2   THEORETICAL FRAMEWORK

It is important to view SM activities through a theoretical lens for various reasons. Firstly, theories can explain the complex and dynamic relationships prevalent in SM. Secondly, theories can help researchers and scholars make sense of demographic, attitudinal, and behavioral data of users. Finally, a solid understanding of theory can help researchers predict future trends and discover new opportunities. A variety of theories can be applied to the understanding of SM networks, information exchange, expression of opinion, the structure of senders and receivers, and content generation and consumption.

2.1 McLuhan's—The Medium is the Message

Marshall McLuhan's idea that the media as opposed to the content is more powerful (McLuhan, 1995). Applying his theory to the social media landscape, one can argue that the modality of this new media in terms of frequency, interactivity, and synchronous nature (or lack thereof) can transform human communication. While the content of the messages are likely to mimic traditional communication, the manner in which communication occurs now deeply impacts business to stakeholder communication. Businesses need to change their advertising, promotional, and customer service behavior due to the usage of new SM services and adopt a positive attitude toward this media.

2.2 Social Network Theory

When it comes to successful SM marketing, Social Network theory is beneficial in understanding the phenomenon. Social Network Theory is the study of how people, organizations or groups interact with others inside their network. One needs to understand the largest unit (networks) as well as the smallest unit which is the actors. Actors consume SM through interactions with each other. They create social ties by creating communities, sharing information units, and finding links between each other. These links could be strong such as close friends or weak such as friends of friends or acquaintances. This is where the strength of SM lies—the potential linkages that can be created via individuals and their followers. SM platforms exhibit varying degrees of creating linkages. While blogs offer weak tie strengths and a one-to-many model of communication, the Instagram network offers strong tie strengths and many-to-many, one-to-one communication model. Applying this concept, businesses should use media networks that offer strong tie strengths and a many-to-many, one-to-one communication model (Hoffman & Novak, 1996).

2.3   Social Exchange Theory

The Social Exchange theory posits that human beings are largely motivated to cultivate relationships by evaluating benefits and risks. (Thibaut & Kelley, 1959). Individuals engage in behaviors they find rewarding and avoid behaviors that have too high a cost. Symbols such as approval or prestige can be included in these behaviors. Therefore, actors may communicate on SM if they perceive their actions will increase their reputation and popularity as well as pave the way for other tangible and intangible rewards. Firms can therefore utilize a variety of advertising appeals to elicit purchase behavior of customers based on the Social Exchange Theory. Firms can utilize communication tactics using influencers, mentions, tagging fellow netizens etc. that allow consumers to enjoy benefits such as exposure and fame within their social circle.

While these theories are not a comprehensive list explaining SM behavior, they provide a useful insight into understanding why individuals use SM, and how firms can capitalize on individual behavior to further their business goals.

2.4   Study on SM Usage by Emirati Entrepreneurs

In order to provide context to the recommendations, the researchers explored how SM was utilized in the overall marketing strategy by entrepreneurs in the UAE. This study situated SM best practices informed by theory within the UAE context by exploring how small and medium firms use SM in their business agenda. In an interview with eight entrepreneurs from the UAE, they elicited responses on how they utilize SM for their businesses. Using a snowball sampling method, two socially active entrepreneurs were recruited who in turn aided in recruiting more entrepreneurs. The responses were transcribed and analyzed using the coding and categorization technique. The following recurrent themes emerged, consistent with global business and media trends and extant literature (Chatterjee & Kar, 2020; Wally & Koshy, 2014).

1. Enhance customer service
2. Increase engagement specially during stressful periods such as the Covid-19 pandemic
3. Managing reputation
4. Getting 'free' publicity

5. Collaborating with thought leaders and influencers
6. General growth
7. Understanding customer behavior and what 'they need'.

In the next section, the current practices were explored and the themes emergent from the interviews were expanded upon as elaborating upon the transcriptions is beyond the scope of the authors' work.

3   Discussion and Analysis

3.1   Use of Social Media in Business: Current Practices

In the modern-day business environment, SM platforms significantly contribute as 'information brokers' and helping to continuously propagate information, connect different channels, create awareness, and maintain interest in a product or service (Stephen & Galak, 2009). Undoubtedly SM is changing the way people approach business. Every businessperson must now have a personal SM strategy to be successful in the competitive market (Dutta, 2010). According to a study conducted by https://www.emarket.com/, 75.3% US businesses will use Instagram, the SM platform, by 2020. On Instagram alone, 80% of the account holders are now following at least one company (Newberry, 2019). The business community can't ignore the power of SM anymore. Many of the business strategists already recognized the potential of SM and using it for different business purposes such as—marketing, reputation management, crowdsourcing, improving customer service, and employee relations.

3.2   Marketing

Marketing heavily dependent on context and external factors. Changes in context and external factors can significantly impact the marketing landscape (Brady et al., 2002). One such factor is the rise of SM. The proliferation of SM in the last decade has not only brought the revolutionary changes in the way people communicate, but also significantly influenced their behavior. Hence, when business firms and strategists use SM, they focus on different aspects of integrated marketing communication and consumer behavior. It includes:

3.2.1   Generating Traffic

Today, when the internet is witnessing an information explosion, getting the attention of the user or target market to the official website of the company is challenging. The user may not notice the details of the product or service posted on the website as one does not follow the website on a regular basis. Here, SM plays a crucial role. Sharing great content or new details about a product or service from the business blog or website to a social channel is a great way to reach the target audience. When posted in SM handle, there is a high possibility of getting the attention of the SM users, which may include the target audience of the business. SM is used in marketing for increasing the visibility of the product or services, for getting attention from new people, for showcasing the expertise, and for driving traffic to the company website. SM platforms like Instagram, Facebook, and Twitter offer advertising formats specifically designed to collect leads. Through sponsored posts, interactive nature of the content, companies can get the users' initial attention. It eventually drives the traffic to the official website of the company and helps in lead generation.

3.2.2   Building the Brand

Brand is the breath of modern business. Brands are not built overnight. SM provides a great opportunity for building a successful brand. Through Facebook pages, Instagram business profiles, Whatsapp statuses or pinned tweets; companies can showcase their product or service and connect with the customers. According to a study, 60% of the Instagram users say they discover new products through the SM platform. In addition to advertising, SM also provides an opportunity to present the human side of the company to potential customers. From extending reach to generate greater brand affinity, SM plays a significant role in every phase of business. Figure 1 shows how SM can influence a user to turn into a brand advocate.

Every SM user's journey generally starts with the basic intention of networking with friends or community. If a company wants to attract the attention of this user, it has to post interesting content, which can catch his attention and convert the user into the viewer of the post. For some time, the user will be just the viewer. However, if interesting content are posted regularly, he will be a fan of the content and of the brand or company posted it. In the next stage, his liking and familiarity with the brand through SM content may prompt him to make his first purchase

Fig. 1 Different phases of progression of a SM user to brand advocate (*Source* Prepared by the authors)

and to be a customer. When he makes the first purchase of the product, if he is satisfied- he will be a repeat customer. Several repeated purchases will make him a brand loyalist. In these brand loyalists, some of them who are vocal may act as brand advocates in SM. Brand advocates proactively talk up, share positive sentiments about a brand in their social networking community. This can be more credible than the company led advertising campaigns. Considering this, now companies aim to increase the number of brand advocates in SM.

3.2.3   Content Creation and Distribution

Web 2.0 has revolutionized the concept of the marketing mix. While the use of websites is playing a critical role in redefining the concept of place and physical evidence; SM platforms are enabling to create new strategies for promotion. SM is an effective channel to present promotional content for potential customers. When people start liking, commenting, and sharing social posts, the content is exposed to new audience-their friends and followers, which will help to get the attention of a larger group of audience.

A research study conducted by Microsoft suggests that the average attention span of humans has fallen since the start of the century. The average human attention span in 2000 was 12 seconds, but by 2013 it reduced to only 8 seconds. SM is widely used by marketing campaigners to tailor the promotional content to this new reality of human attention. Short promotional videos of 6–30 seconds are gaining popularity in the new media world. Besides, SM enhances and encourages interactive activities. Through likes, comments, and shares of SM, marketing content can go viral. This is a very cost-effective way of content distribution. SM has shown a new way for content creation and distribution in marketing.

3.2.4   Word of Mouth Through Social Marketing

According to the Nielsen's Global Trust in Advertising Report issued in 2012, consumers are more likely to trust recommendations from known sources as opposed to advertisements and commercials (Grimes, 2012). However, the frequency of people meeting their family or neighbors in person is coming down due to advancements in communication technologies. And, during the COVID-19 pandemic around the world, people were kept on lockdown. It forced them to accelerate the digital usage and virtual friendships on SM. This has increased the scope for digital media word-of-mouth marketing. Traditionally *word of mouth implies* a message passing from person to person using oral communication. In the digital age, in SM the recipients are in charge of generating word of mouth. The word of mouth gets attention and creates a buzz about a product. Since the worldwide web doesn't have geographical boundaries, word of mouth can reach every corner of the world. Many purchase decisions are driven by word-of-mouth advertising such as consumer reviews, blog posts, influencer usage etc. (Bughin et al., 2010). Now marketers now can plug-in to the real, uncultivated, word-of-mouth conversations of the consumer through SM, making the reviews more credible. Consumer concerns, comments, questions, opinions, likes, dislikes which are available in the public domain are a goldmine of information. It can provide new insights into multi-dimensions of business.

3.2.5 Influencer Marketing
Traditionally, a business collaborates with an influential person to promote a product or service. Celebrities are used mainly for TV, print, and outdoor advertisement campaigns. In the digital world, using SM, the common man is acting as a celebrity. The common people who have dedicated and engaged groups of followers on SM are emerging as the brand ambassadors of companies. They are known simply as 'SM influencers'. According to the Advertising Authority of the UK, if someone has more than 30 k followers, he can influence the purchase decision-making behavior of his online followers. Influencer marketing is emerging as the most powerful strategies of digital marketing. Almost half of US and UK digital marketers spend at least 10% of their marketing budget on influencer marketing.

3.2.6 Targeted Advertising
The new-age business environment is about 'competing on data' and 'competing on analytics' (Davenport, 2013). The traditional business of data collection as the starting point for advertising research insights is changing because of the research and developments in the field of big data analysis tools and AI (Artificial Intelligence). Gathering relevant SM data, analyzing it using advanced tools helps to recognize the problem, detect the opportunity, generate insight and target the right people through SM advertisements strategies. With the emergence of SM tracking and the use of advanced analytics tools, it is easy to guage the full impact of SM marketing activities in real time. Analytics tools like Awario, Snaplytics, Keyhole,Union Metrics, Rival IQ, Squarelovin, Sprout Social,

and Hootsuite are helping the business community to track website traffic generated from SM, conversions and email sign-ups. These tools enable marketers to precisely target potential customers with very little cost. In a traditional media setup, it is always a challenge for marketers to prove return on investment (ROI) on the advertising budget. However, ROI for both organic and paid SM ad campaigns can be easily calculated. From startups to established companies'—use of SM has become integral part of advertising and marketing strategies. Therefore, in the US companies have started to spend more advertising budget on the Facebook ad than all print ads put together. In the UAE, due to the high internet penetration in the country, social media advertising is emerging as the most preferred form of advertising for the business community.

Beating Competition: Understanding what people are saying about competitors is also important in marketing. Tracking the consumer response of a competitor brand using traditional media is challenging. SM enables to overcome it. The big data analytics tools help to track mentions, impressions, comments, shares, likes, and dislikes of competitor brands. The finding from the data analysis can reveal the pain points of the rival brand. If a company can address the issues found and reach out to the aggrieved consumers of its competitor brand, winning new customers is an easy task for a company.

3.3   Reputation Management

SM is the most democratic media platform available today. People express their views without any fear. The positive or negative opinions expressed by the people can make or break a brand and company. In a survey conducted in 2019 by Bright Local Consumer, it was discovered that 82% of consumers read online reviews even for local businesses. The number of consumers who 'always' read online reviews when they make purchase decisions are now at 36%, up from 27% in 2018. One negative review, response or SM message may leave a bad impression in the minds of the consumer. Hence companies are trying to focus now on reputation management using SM.

3.3.1   Resolve Issues Affecting Reputation

Social networks have leveled the playing field by opening up the communication channels between brands and customers. Now customers can interact directly with brands and have a conversation in real time. Any issues affecting the customer and the brand reputation can be resolved easily. If someone is posting a bad experience, the company can provide clarity, resolution to the problem or at least issue an apology.

3.3.2   Handling Negativity

It's important to not avoid customer feedback posted on SM. Both positive and negative comments should be addressed. Even an aggrieved customer should get a polite response in SM. This, in addition to showcasing the willingness to accommodate criticism, enhances the brand image. It also conveys a larger message to the potential consumers that the company is customer-centric and treats every feedback of the customer with the utmost respect. The customer is king in business. When he is treated respectfully, brand reputation grows.

3.4   Customer Management

Using SM for customer interactions, specifically customer service is no longer a recommendation. Rather it is a requirement. Customers expect brands to be available on SM and often reach out to brands on SM. Evidence demonstrates that more and more consumers used a company's SM site for servicing. A JD Power study indicates that 67% of consumers use social media for customer service interactions. This finding is especially true for millennials (J.D. Power and Associates 2013 SM Benchmark Study). Consumer expectations from brands has now moved from customer service to customer care. It is important to note that customer care is a more pro-active form of customer management where brands reach out to customers even before making a purchase. This could entail *listening* in to identify potential spaces on SM—be it sending informational messages to customers; offering service to potential customers etc.

Thus, as more consumers seek the social space for customer service, it is imperative for brands to be present and pro-active on SM as well. Mere presence does not qualify, brands now have to listen and respond to consumers strategically. Brands who ignore customer service requests on SM see an average churn rate that's 15% higher than companies who don't. Additionally, customer service interactions are in public domain for everyone to see, therefore lack of proper customer service may have a detrimental effect on potential customer acquisition.

According to a study conducted by Bain and Company, if brands engage with its customers and responds to their service through SM, it will have positive effects. Those customers may spend 20–40% more on the products and services of that brand. Another study conducted by HBR, which was done to understand consumer behavior in Twitter, found that when costumers get a response from a brand, especially within the five minutes, chances of repeat purchase are very high and time they spend to study the brand is more. In addition to this, study also found out that these customers recommend the brand and products to others as well (Huang et al., 2018). Research studies show that even the intent shown by the brand to solve the issues of a consumer in SM is appreciated. Even if the complaint is unresolved, an acknowledgment of the customers issue can create a sense of reassurance about the brand. The trust earned through this action can influence their purchase decision as well (Kim & Ko, 2010).

While companies may maintain SM accounts or other customer service teams, there is a need to coordinate their efforts. Successful companies tend to invest more on integrated SM customer service strategies with efforts linked to their business goals. However, a lot of brands simply exist on SM and perceive to be sorted. Below we list some best practices that brands may employ toward their customer service efforts.

3.4.1   Join Social Media

If a brand is not on SM, then joining a platform is necessary. Customers are on SM and brands must take advantage of their presence. In fact, figuring out which particular SM networks are being used by customers is essential. While Twitter is the preferred network for issues, Pinterest may be optimum if you sell products. It is vital to establish where your audience is.

3.4.2   Respond Promptly

SM customer support creates an 'always available' expectation and customers expect a response within minutes of a complaint. Therefore, responding to inquiries, reviews, and complaints as quickly as possible is the best SM customer service strategy. Poor or no response to SM queries can cause people to spread negative word of mouth (WOM) regarding the brand. Negatively balanced WOM translates to customers buying less from a company in the future; not recommending the brand's products or services to others; using another channel to escalate concerns; complain publicly via SM; posting a negative review online, etc. In short, businesses may lose billions of dollars due to poor customer service.

3.4.3   Respond Directly and Genuinely

There is immense value in making a personal connection with a customer. Using their name or personalizing the response can go a long way in terms of humanizing a brand. In fact, customers are willing to pay more if their name or initials are added in the response. Similarly, if the customer service agent signs off with his/her name it can further humanize the brand. In addition to this, customers will feel more comfortable in later stages of follow up, if they know name of the employee who helped them first.

3.4.4   Respond to Positive Comments

While negative comments need to be addressed, a positive review needs acknowledgment as well. The key to turn a one-time customer into a loyal one is to focus on engagement and thank them publicly. Every post, review, and check-in on SM needs acknowledgment—just like an in-store customer would not be ignored by any employee.

3.4.5   Take Ownership of Mistakes

A mistake needs to be acknowledged. The best practice for a brand would be to show transparency, honesty, and a willingness to fix the situation right away. It is also important to refrain from negative language even when the customer is angry. Millions of viewers will be able to read comments, so it is vital to use positive language and apologize when required.

3.4.6   Decide Where to Solve the Problem

Some comments are best addressed in the public eye. However, certain communications need to be taken offline—it is vital to remember that SM is a community-based platform and it can prompt the community to join in and create an unwanted situation. Direct Messages, emails, and phone calls can work in such cases. The key is to demonstrate a willingness to respond to the customer.

3.4.7   Use Monitoring Tools to Aid SM Customer Service

It is now easy to monitor customer service activities in SM with tools. The advanced technological tools enable to listen to customers in multiple

SM platforms. All these accounts can be integrated, and a single customer service representative or team can monitor and manage all the accounts from one place. Examples of such tools are—Hootsuite, Sproutsocial, Buffer, etc. These tools continually add new features and updates to their platforms that make it easy for brands to deliver the best service possible to their customers on SM.

3.5   Crowdsourcing

Unlike before, today in a technologically empowered world; the consumer has the tools, opportunity, and desire to be involved in business activity on a much deeper level. Hence, crowdsourcing has been very successful. It helps to bring people together online for collective, thoughtful engagement and get involved in the production and promotion processes of a business. With the help of SM, through crowdsourcing, businesses bring consumers together and closer. It engages them through interactions and tries to build relations with a brand. By asking for input and engaging them business several activities it can create a personal connection, which in turn leads to greater brand loyalty (Spiegel, 2011).

3.5.1   Engaging with Consumers
Research show that engaging with customers and followers can give a rich dividend in business. It helps to stay active and respond to comments and questions on the company's SM posts in a way that's appropriate to the brand. Instead of becoming overly self-promotional, companies can engage with costumers and use their positive experiences and comments for the marketing of the brand. User-generated content adds more reliability to a brand than the official SM marketing content.

3.5.2   Polls and Contests
By getting feedback about a product or the service, the gap between satisfied and disappointed customers can be easily reduced. It creates a feeling among the dissatisfied consumers that the company is listening to them. Today, polls and surveys administered through SM are helping to get feedback from a larger group of consumers. Along with the polls, contests are also used as crowdsourcing activities. Contests in SM can generate positive feelings about a brand with very little cost. For example, Starbucks in 2014, arranged a 'white cup contest'. The contest was arranged in SM platforms, where the customers of Starbucks were asked to design

their Starbucks cup with their own design, take a photo, and upload it to SM using the hashtag #WhiteCupContest.

Over a few months, more than 4000 people submitted their designs to the contest. Contests like these, help to engage the consumer with the brand and to increase its visibility (Speier, 2016).

3.5.3   Crowd Funding

Ideas are generated exponentially in a connected world, with intellectual diversity and wealth of knowledge available at the common man's disposal. But these ideas need financial support to see the light of the day. SM using the crowd-funding technique has given a new venue for sourcing the required funds to the common man. Crowdfunding, in the digital age, leverages the advantages of online platforms and SM to help people raise funds for common man's ideas, business or charity. Using SM people can pitch in their ideas or business plans along with a fund-raising goal easily. And SM users respond to it.

3.6   Employee Relations

SM is a vital form of communication for businesses as they can take advantage of this medium for external stakeholders. The same communication benefits can be reaped for internal stakeholders—employees, as well. Businesses that implement an internal communication platform see many benefits, both in terms of profits and employee satisfaction. While employee engagement remains a priority for communication practitioners working in corporations, only 33% of U.S. employees and 15% of global employees are engaged at work (Boyle, 2017). Notably, companies have massively incorporated SM to improve corporate reputations, and employee engagement should be seen as part of this. But many organizations are still not taking the advantages of the changing technology and tools to the full potential. The skepticism arises primarily when SM is used to facilitate employee collaboration and knowledge-management, but optimal results are not achieved (Wehner et al., 2017).

According to a study conducted by Winograd and Hais (2014) for Brookings report, millennials will comprise more than 30% of Americans by 2020 and it will increase to 75% of the workforce by 2025. Most of these millennials spent the majority of their lives with a SM presence and they expect communications to be two way and engaging. Studies show that engaged employees are always a competitive advantage, regardless of their generation.

3.6.1   Effective Communication

Social channels allow employers to communicate information publicly and privately. Regular organizational communications, team project information, meetings etc. can be managed via SM platforms such as Facebook and Yammer. Such sharing of information come can help cut down on the daily deluge of email many find in the workplace. Google and Starbucks use Twitter to engage and inspire employees by sharing short bursts of information. Similarly, The Massachusetts Institute of Technology, through its @MITWorkLife handle, promotes work-life content and professional development events. Additionally, training of employees can be negotiated over SM with platforms such as Periscope and YouTube. The latter allows archiving training broadcasts with the additional benefit of making the channel public and thus help in building the brand's reputation. It may be argued that Facebook or other SM platforms may distract employees, and internal SM such as company blogs may help organizations achieve long-term goals, including engaging employees. Blogs also can be used as a tool for internal communication in an organization. It can help the organization to post the details of benefits available to employees, to conduct Q & A sessions, to share employee feedback through surveys, and to publish the announcements related to the daily operations of the organization.

3.6.2   Encourage Thought Leadership and Brand Promotion

Employees are the best ambassadors for a brand, and SM amplifies their voice, providing them an opportunity to promote a product or an industry-related thought. When employees share or respond to a company post, they expand the reach of the company by widening the audience that has the potential of generating leads and converting leads into customers. This allows them to be thought leaders—individuals who create, share, and comment on educational content that is relevant to their position or the organization's industry. Such content is often interesting, newsworthy, and innovative information that customers would not have likely discovered on their own. Thought leadership provides new ideas, perspectives, and data that teach and shift your audience from their status quo. This content often comes from a trusted source and can create a dedicated following or audience who are interested in action on new ideas." While top executives of a company frequently engage in thought leadership, the average employee is an important resource that needs to be tapped into as they engage with customers, prospects, and potential future employees on a daily basis, thus significantly impacting the public perception of the brand.

3.6.3 Foster Creativity, Internal Expertise, Corporate Culture

SM networks allow employees to freely share questions and answers and showcase their own knowledge. Blogs and forums allow for such sharing of expertise through their conversational and informal nature—employees can simply toss out ideas. Employees can also place themselves as experts on an area by posting content. Furthermore, employees can use blogs to personalize relationships with other employees and members of other strategic publics (customers, stockholders, etc.). On an enterprise level, internal SM channels can empower employees to engage with each other. Channels like Beekeeper allows for bottom-up communication in ways that have not been possible before. After a merger between Fairmont, Waffles, and Swissotel hotel groups, internal communication reflected true Swiss heritage—democratic. By allowing employees to connect with each other from around the world—especially the class of employees that may not even have a company email address is deeply empowering, and profoundly affects company culture. Employees can share their insights, ideas and discuss issues related to professional development and work experience with each other. Internal SM means having a common digital platform, which is accessible to everyone around the world—a maid in Russia can now comment on a post from a concierge in Switzerland. It also allows for storytelling, thus shaping culture by sharing celebrations, stories of excellence, every-day work experiences, and community involvement.

3.6.4 Employee Recognition

Employees like public recognition. SM provides opportunity to show case their expertise, talent and to get recognition, which eventually helps to create personal branding. Sharing of accomplishments, 'employee of the month' recognitions can encourage high performance from employees. Creating content that highlights excellent work via videos is yet another way to empower the employee by allowing them to amplify such content.

Gamification is also a trend that is catching up in the social world—a fun competition in a sales profile may sound like bragging rights but is an effective method for incentivizing performance. The key here would be to celebrate not just the same people over and over again, but to highlight others with honors such as 'most improved', 'A for Effort', etc. Internal SM platforms may also get the attention of employees by offering them free stuff in exchange for platform participation. A major grocery store chain has enabled employees to redeem points accrued from platform activity for company merchandise. Additionally, there is immense value in posting spontaneous and candid content over well-planned content—unscripted moments can recognize and celebrate employees in a natural fashion. Encouraging employees to post their own photos at fun company events like picnics, volunteer activities, and even company town halls can go a long way. Finally, it is important to amplify the content of employees by retweeting or reposting their tweets and shares as examples for the rest of the organization to follow.

3.7   Future Challenges and Opportunities

3.7.1   Developing a SM Strategy

Creating a SM strategy sounds primitive for this chapter, but many companies simply exist on SM without a definitive strategy. An SM strategy will help a brand negotiate its goals. It is important for brands to align their goals with their broader marketing strategy and mission. The strategy will also determine the success and failure of the brand's campaign. Specifically, SM audits are vital for brands as they help understand what is working, what's not and what can be improved upon. Audits should be repeated regularly in order to improve performance. Furthermore, it is imperative to select the correct metrics to gage performance—one has to move beyond superficial followers and likes and focus on leads, link clicks and impressions. Facebook and Instagram have tested the removal of likes and this may signal a shift away from traditional and superficial data. Finally, social listening will be crucial in the coming years. What is being said on SM will hold more value than how many people view and like a post—conversational and sentiment analyses will gain importance—therefore brands must focus on these metrics.

3.7.2   Humanizing the Relationship with the Audience

SM helps build a brand and attract new customers, but retaining the customers is also important. Brand loyalty is important and enhancing engagement and developing a long-term humanistic relationship is vital.

While there are many ways of doing this, most outside the scope of this article, two trends that can humanize a brand should be explored as much as possible—User-generated content (UGC) and storytelling. While UGC is not something novel, it will continue to enjoy an upward trend with more and more brands utilizing it. When brands motivate their customers to create content and tag the brand for a chance to get featured on their page, they create authentic content that can help the brand increase credibility and trustworthiness. In the context of the UAE study the authors conducted, humanizing the brand especially during the pandemic reinvigorated the businesses that suffered stagnancy due to the crises. The entrepreneurs reported that by increased engagement they reaped benefits such as over-booking of services post lockdowns.

Storytelling can also help create sticky content that moves beyond brute selling. Through storytelling brands can introduce core values, giving customers relatable reasons to stick around. Dove is one such brand that capitalizes on inspirational and story-driven content that can help humanize brands and instill brand loyalty. By publishing such content, brands represent their missions while making them relatable to their core publics.

3.7.3 Perfecting the Art of Customer Service

Most customers prefer SM to communicate directly with brands. However, the response rate for costumer's query is very low-just 11%. This could be because of the high volume of the messages or queries brands get from costumers on daily basis. There are advanced ICT and big data tools available today which can help to address this issue. Every effort should be made to respond to every SM message. Using SM management tools such as Buffer, Sprout Social, Hootsuite, etc. can ensure that no message goes unnoticed. These tools can identify areas of issue resolution; join the conversation; or even an opportunity to delight and entertain the customer. Adding on to the previous point on humanizing the brand, outstanding customer service is yet another method of doing so. By acting like a human and reaching out to customer messages—be it positive or negative, a brand can easily negotiate and shape its reputation on social. This recommendation is consistent with the findings of the study wherein the entrepreneurs reported opportunities for turning angry customers into loyal ones.

3.7.4   Capitalizing on Video Content

Video is a megatrend—brands are driving tremendous growth with video across SM platforms. In fact, long and short form videos are among the most shared content on SM. Views of branded video content have increased 99% on YouTube and 258% on Facebook. Moreover, 64% of consumers believe that marketing video posted in Facebook has influences their purchasing decision (O'Neill, 2017). Brands cannot ignore the power of video—be it ephemeral content via stories, IGTV, and the increasingly popular TikTok—they should seize the opportunity to repurpose video content from other platforms to increase brand awareness and create loyalty. Entrepreneurs in the UAE have reported massive potential in promoting their brand using educational, entertaining, and convincing video content.

3.7.5   Integrating Artificial Intelligence

An area of opportunity in the future is the integration of AI into many marketing efforts—as business encounter massive competition and a pressure to excel, they need to scale up. There are several ways brands can integrate AI in SM strategy in the future. Conversational AI can be used to automatically generate responses for users' requests on SM. This can remedy the problem arising with poor, delayed, or no customer service. Chatbots are one such interface that can engage with human clients—it is at a basic level as of now, but this is an area that should be explored by brands. Test runs and studies have demonstrated that chatbot interfaces may be as good as human agents in showing empathy to help users cope with emotional situations (Xu et al., 2017). That said, the major challenge will be in creating synergy between AI and human interaction systems in a manner that retains and celebrates the human and personal aspect of the brand.

AI can also help in creating social content. Although possessing limited capabilities currently, AI subsets such as natural language generation can be trained to create high impact reports or fact sheets for brands. Similarly, AI can help to collect data from millions of posts, at scale. These data not only help to gain insights on current trends, but also for making future predictions and decisions. Other associated immersive AI technologies such as Virtual Reality (VR) and Augmented Reality (AR) can drive future digital markets. Brands like Amazon and Sephora are already exploring AR to provide the customer with a better shopping experience. Sephora, for instance allows their customers to try on makeup before purchasing via Facebook Messenger, using AR filters. This gives shoppers a unique experience and helps them make a better purchase decision. In the future, brands should look forward to leverage various other AR functionalities.

3.7.6 Targeting Nano and Niche

SM marketing will see a more fine-tuned and precise audience identification approach. Ads can be created to appear on the wall of one person. Termed NanoAd—it may not sound profitable, but a brand may chance upon multiple avenues if it targets the CEO of a large company or a venture capitalist. Brands must engage in active social listening once they find their target audience to explore their needs and service requirements and engage with them. Such engagement may allow audiences to directly purchase goods and services from brands. SM platforms should continually roll out new features to deliver customized ad experiences to every individual. Currently, Facebook uses its machine learning algorithms to collect data and other signals from its platforms. It combines this data with the comments and insights shared by consumers to decide the right people and predict their behavior for a given message. For example, many consumers prefer one format of shopping over another. They are likely to engage with the content in when it is presented in their preferred format. This allows a brand to present their ads in different formats. Facebook can help to improve the engagement with the ad content, as it has the big data base of information on its users preference. Facebook Dynamic Ad unit can enable brands to create ads for the catalog sales, improve traffic and also for the conversions. In order to target specific audiences, brands should consider testing and incorporating solutions into their SM marketing strategies going forward.

3.7.7 Focusing on Organic Reach

A steady decline in organic reach is on the rise. Attention is getting costlier. Brands will need to seize every opportunity to fight this in the coming years. For businesses, low level of reach translates to a very low return on investment and slim conversions. It is vital for brands to create quality content in order to drive reach. The right content mix for audiences and the use of analytics to notice performance trends is non-negotiable moving forward and brands must capitalize on these to fight the decline in organic reach. SM platforms use algorithms, and it is easy for brands to blame them, but brands will need to build a solid following and concentrate on engaging and serving current followers so more organic reach will occur. Content should always be convincing, educational, and entertaining. Finally, posts will need to be consistently published.

3.7.8 *Thriving in a Regulated Environment*

SM marketers will face strict regulation in the coming years and will find it challenging to target their audience. The impact of regulation can be seen already—Facebook will soon require marketers to confirm they have user consent for Custom Audiences. Furthermore, issues of privacy and storage of first-party data will invade the world. However, it may be possible to negotiate collaborative spaces between targeting and regulation. Looking at offline behavior of the audience may be a solution—surely other solutions may be explored. A successful brand in 2020 and beyond will move beyond audience targeting and view SM targeting as one part of the marketing mix and use regulations to their benefit, not viewing them as a limiting factor.

4   Conclusion

Brands know that being on social is the default experience for customers. However, few brands have moved beyond optimization. It is hoped that businesses will leverage some of the trends that will dominate the SM landscape in the next decade. It is vital for brands to keep abreast of latest developments as the SM landscape is dynamic and extremely competitive. Brands should not just focus on SM as a platform for marketing, rather view it as a place to build trust and maintain rich engagement with their audience. Moving forward, brands will need to look at emerging markets to prepare for the next shifts in mobile technology—reaching the 'next billion' users is important and non-negotiable. Drilling down to regions, using localized and vernacular content may be the order of the day when it comes to reaching these emerging markets. Such SM strategies will enable brands to create a genuine and relevant experience for audiences; and help them stay ahead of the curve. The context for the study was UAE, specially selected due to its position with remarkable internet and SM usage. SM being the default media platform for business activities. Combined with a generation of young entrepreneurs running their own businesses, SM is bound to revolutionize the future of the Middle East and North Africa (MENA) region and many other emerging markets in Asia.

REFERENCES

Brady, M., Saren, M., & Tzokas, N. (2002). *Integrating information technology into marketing practice—the IT reality of contemporary marketing practice.* Academic Press.

Boyle, E. (2017, December 13). *How HR leaders can win a seat the table.* https://www.gallup.com/workplace/231644/leaders-win-seattable.aspx?g_s ource=EMPLOYEE_ENGAGEMENT&g_medium=topic&g_campaign=tiles.

Bughin, J., Doogan, J., & Vetvik, O. J. (2010). *A new way to measure word-of-mouth marketing.* Mckinsey & Company. https://www.mckinsey.com/business-functions/marketing-and-sales/our-insights/a-new-way-to-measure-word-of-mouth-marketing.

Chatterjee, S., & Kar, A. K. (2020). Why do small and medium enterprises use SM marketing and what is the impact: Empirical insights from India. *International Journal of Information Management, 53.* https://doi.org/10.1016/j.ijinfomgt.2020.102103.

Davenport, T. H. (2013). Analytics 3.0. *Harvard Business Review.*

Dutta, S. (2010). Managing Yourself: What's Your Personal Social Media Strategy? *Harvard Business Review, 88*(11), 127–130.

Hoffman, D. L., & Novak, T. P. (1996). Marketing in hypermedia computer-mediated environments: Conceptual foundations. *The Journal of Marketing,* 50–68.

Huang, R., Ha, S., & Kim, S. H. (2018). Narrative persuasion in social media: An empirical study of luxury brand advertising. *Journal of Research in Interactive Marketing, 12*(3), 274–292. https://doi.org/10.1108/JRIM-07-2017-0059.

Graham, M. (2018). Customer service through SM. *Printing Industries of America, The Magazine; Sewickly, 10*(5), 18–19.

Grimes, M. (2012, October 4). Global consumers' trust in 'Earned' advertising grows in importance. http://nielson.com/. http://www.nielson.com/us/en/press-room/2012/nielson-global-consumers-trust-in-earnedadvertising-grows.html.

Kim, A. J., & Ko, E. (2010). Impacts of luxury fashion brand's SM marketing on customer relationship and purchase intention. *Journal of Global Fashion Marketing, 1*(3), 164–171.

McLuhan, M. (1995). *Understanding media: The extensions of man.* Cambridge.

Newberry, C. (2019, October 22). 37 Instagram stats that matter to marketers in 2020. https://www.hootsuite.com/. https://blog.hootsuite.com/instagram-statistics/.

O'Neill, M. (2017, Jun 6). The State of Social Video: Marketing in a Video-First World [Infographic].

Stephen, A. T., & Galak, J. (2009). The complementary roles of traditional and SM in driving marketing performance. *Insead Working Papers Collection*, 1–38.

Speier, K . (2016, January 7). 4 examples of clever crowdsourcing. http://mai nstreethost.com/. https://www.mainstreethost.com/blog/four-examples-of-clever-crowdsourcing-campaigns/.

Spiegel, R. (2011, June 16). *3 Ways to benefit from SM*. https://www.socialmed iaexaminer.com/3-ways-to-do-social-media-crowdsourcing/.

Thibaut, J. W., & Kelley, H. H. (1959). *The social psychology of groups*. Wiley.

UAE SM Usage Statistics. (2020, April 29). *Global media insight*. https://www.globalmediainsight.com/blog/uae-social-media-statistics/.

Wally, E., & Koshy, S. (2014). *The use of Instagram as a marketing tool by Emirati female entrepreneurs: An exploratory study* (pp. 1–19). https://ro.uow.edu.au/dubaipapers/621.

Wehner, B., Ritter, C., & Leist, S. (2017). Enterprise social networks: A literature review and research agenda. *Computer Networks, 114*, 125–142. https://doi.org/10.1016/j.comnet.2016.09.001.

Winograd, M., & Hais, M. (2014). *How Millennials could upend Wall street and corporate America* (pp. 2–4). https://www.brookings.edu/wpontent/upl oads/2016/06/Brookings_Winogradfinal.pdf.

Xu, A., Liu, Z., Guo, Y., Sinha, V., & Akkiraju, R. (2017). A new chatbot for customer service on SM. In *Proceedings of the 2017 CHI Conference on Human Factors in Computing Systems—CHI 17*, 3506–3510. https://doi.org/10.1145/3025453.3025496.

The Prospects and Risks of Industry 4.0: Issues and Implications

Mythili Kolluru and Shobhna Gupta

1 Introduction

In the year 2011, the concept of Industry 4.0 was initiated to provide a suitable platform for the progress of German economic policies (Mosconi, 2015) with a new insight. In 2015, the Executive Chairman of World Economic Forum, Klaus Schwab (2015), introduced the phrase "Fourth Industrial Revolution". The theme of the World Economic Forum meeting (2016) in Davos, Switzerland, was "Mastering the fourth industrial revolution" (Schwab, 2016; World Economic Forum, 2016). It discussed how the Fourth Industrial Revolution had impacted the current and emerging environment significantly. It also analyzed the way human life and work is affected using disruptive technology such as Virtual reality (VR), the Internet of things (IOT), Robotics, and Artificial intelligence (AI).

M. Kolluru (✉) · S. Gupta
Professional Studies and Undergraduate Department, College of
Banking and Financial Studies, Muscat, Sultanate of Oman
e-mail: mythili@cbfs.edu.om

S. Gupta
e-mail: shobhna@cbfs.edu.om

N. R. Al Mawali et al. (eds.), *Fourth Industrial Revolution and Business
Dynamics*, https://doi.org/10.1007/978-981-16-3250-1_11

223

The difference between natural and artificial is slowly blurring the lines between the physical, biological, and digital systems, leading to an unprecedented fusion of these domains. One of the main features of this revolution is that it will not change what we do but how we think, behave, and change us also to some extent. It brings us the potential for a new renaissance that can transform our universe and our relationships at work, life, and the universe. Lessons from history teach us that recent dynamic changes initiate value shifts. Industry 4.0 has opened an unprecedented change in humans' way of living. It has opened channels of creativity and innovation that have led humans to the path of progress. The leveraging of asset tracking, IT, 3D printing, robotics, the internet of things, the cloud, big data, and sensors have enabled economics to tackle and decouple growth from resource constraints. The impact of Industry 4.0 is enormous. It is affecting cities and design. For example, a world without plastic will have an exponential effect on economies and climate change. We are inspired by nature, but we can create various products and parts with innovative natural ingredients.

Today, we differ not only in the material aspect, but we are changing the biological functionality. It fuels us toward having clean air, water, soil, land, and energy for all generations. It is not a cure or fix-it-all solution. Human beings are needed in all areas, but robotics has made huge improvements, especially in the manufacturing sector. The new paradigm demands the need to embrace change and teach it to the younger generation so that they can understand technology's benefits and lead a better life. Digital technology can empower people, change outcomes and steer the world to more equitable inclusive growth. For centuries, we have been using tools and equipment, but now, we have a machine to augment us because of recent technology advancements. Scientific revolutions in varied aspects like genome editing and digital technological innovations have enabled scientists to shift quickly to find solutions for various diseases. It brings customized offerings and improved or new models of business operations. One of the features of Industry 4.0 is the global connectivity of value chains across companies. This investigation aims to analyze documents entitled "Industry 4.0" and related reports available in Google Scholar and other online consultancy sites to provide a clear insight on the subject. The paper offers the views concerning Industry 4.0 concept and its potential benefits and pressing challenges and concludes by presenting some suggestions for future research.

1.1   History of Four Industrial Revolutions

The first industrial revolution was the steam-powered factories, application of science to mass production. Manufacturing was the second industrial revolution, and digitalization was the third industrial revolution. The fourth industrial revolution is the use of technologies such as artificial intelligence, genome, and Robotics, etc (Table 1).

Technologies of the 4th industrial revolution such as 3D printing, artificial intelligence, genome editing, augmented reality, robotics, digitalization, and the internet of things change the way humans exchange, create and distribute values. 4th IR will transform individuals, industries, and institutions as it occurred in the previous revolutions also.

2   Literature Review

Though Industry 4.0 is in its nascent stage, yet it is impacting industries across the globe. The literature review sheds light on the scholarship related to the earlier revolutions, disruptive technologies of industry 4.0, and trends.

In his book, Schwab (2016), "The Fourth Industrial Revolution", provides a detailed description of industry 4.0. He states that the new technologies are going to transform our homes, workspaces, and lives dramatically. The author discusses the challenges and prospects it poses. The author concludes by offering practical suggestions as how to harness the potential of industry 4.0 to develop initiatives and economics in an inclusive approach.

The World Economic Forum (2016) report details the implications of current and future disruptions on employment across different industries

Table 1   Four Industrial Revolutions

Industrial revolution	Characteristics	Era
1st IR	Steam and water power Mechanization	1784
2nd IR	Electricity and Mass production	1870–1914
3rd IR	Automation of IT system and Electronics	Early 1970's
4th IR	Cyber Physical system	Today

Source Compiled by authors

in various countries. The report presents suggestions related to educational systems, pathways to develop the attributes of lifelong learning among the youth, and insights about how people should collaborate to leverage opportunities of industry 4.0. The report concludes by stating how businesses should work with partners to deliver maximum benefits to humanity.

In his work, Butler-Adam (2018) elaborates on the implications of Artificial Intelligence on economic development. The impact on business, industry, and lifestyle has been extensively discussed here. He concludes by emphasizing the need for all society members to understand the implications which new technologies bring in the wake of industry 4.0.

Feinstein (1998) explains the industrial revolutions in general. The author states how revolutions have touched almost every aspect of our daily life. The author concludes by saying that there has been a steady increase in the standard of living. However, some researchers claim that they did not witness the earlier revolution's valuable and meaningful impact until the late twentieth century.

Sommer (2015) articulates the automation and digitization processes, and the use of electronics and information technologies to characterize the main features of the fourth industrial revolution. With the advancement of technologies, such as 3D printing, the growth of online sales services, such as car services, medical tests from home, and obtaining food directly, sent from the store to the refrigerator, will have a notable impact on small and medium-sized enterprises.

Hessman (2013) notes that Industry 4.0 is still in the early stages with respect to usage in manufacturing, individual context, along with experimental research. Importantly, the German governmentis trying to synthesis innovative technology based "smart" factories that can be installed in universities, research institutes and companies. For instance, Hamburg plant, which is a 100,000 plus square foot factory where most of the units are fetching and assembling components without any human input.

Several studies have traced the trend of creating a communication link for unlimited and uninterrupted flow of information in real-time situations for professionals, like-retailers, health care workers, manufacturers, homemakers, co-workers, customers, and power suppliers for flourishing of Industry 4.0 (Kagermann, 2015; Yu et al., 2015). A notable moment in this context is the exchange of information between machines. The aim of such automation is the customer preferred adaptation of products and services, which adds value to customers and companies.

Researchers have indicated that the growth of industry 4.0 is not limited to robotics and automated production because it is about the digitalization of business processes (Kane et al., 2015; Schlechtendahl et al., 2015). The processes will augment certain automaticity of the workers. However, humans still shall need to use their intelligence. The value will be found in new products and new solutions.

Fonseca et al. (2016) elucidated that we are a part of fast paced computerized world with marked features of on the dot access of information and strengthened mobility due to rapid growth in communication and information technologies (ICT). Concurrently, we need to address global subject of concern such as, improving the health standards and measures for environment safeguard and eradication of poverty. Stakeholder's emphasis companies to maintain a holistic clear and accountable attitude to obtain economic and social growth, along with development of sense of awareness and respect for the environment.

Mokyr (1985) alludes that "Industrial revolutions" has changed current prototype due to alteration over time, in terms of complexity and productivity. Industrial revolution aims to bring forth enhanced productivity due to implementation of innovative technology, thereby causing a significant influence on social, economic and environmental dimensions. Industrial revolution leads to society's modification in a concrete and pragmatic manner irrespective of supporting scientific advancement.

Schlick, Stephan, and Zuhkle (2012) document the background of industry 4.0, which is the first industrial revolution highlighting the inauguration of power loom in 1784. This revolution emphasized the mechanical method of goods production, and transportation. Second Industrial revolution is marked by—Electrification and Industrialization. Advent of ICT, computers along with programmable logic controller in 1969, marked the emergence of automated era known as the third industrial revolution. With the predominance of electronics in production processes in this time period it is often termed as electronic automation.

According to Vogel-Heuser and Hess (2016) German economy was uplifted with the entry of Industry 4.0 in year 2011. On similar lines, EU and its associated countries have taken the initiative to develop industry 4.0. Moreover, China also launched China Manufacturing 2025 (CM2025), which is in accordance with industry 4.0.

An industry 4.0 is based on human machine interaction, along with usage of cyber-physical systems for processing complex data. The various robust features that bring versatility are highly customized and digitized automated platform for production. Furthermore, effective communication along with automatic data exchange and value-added services are characteristic attributes, Posada et al. (2015).

Lasi et al. (2014) have particularized on discoveries and inventions related to technologyin the remodeling of enterprises' performance. Some examples include the internet, social networks, apps, smartphones, systems engineering, laptops, 3D-printers, machine learning, and artificial intelligence. The development of these technologies can further improve production and supply of goods in terms of automation and digitalization. Moreover this would usher in new changes in the underlying assumptions of the technology itself and shall transform the industry.

Researchers have recognized that (ICT) is the connecting link to join both physical and virtual worlds known as cyber-physical production systems (CPPSs). The later consists of integrated network system consisting of social machines linked online along with mechanical and electronic components (Kagermann & Helbig, 2013; Lasi et al., 2014). Industry 4.0 improvises production system and business models and thereby strengthens both society and environment. The authors strongly emphasize that companies should focus on both production and services. In such a situation, the Industry 4.0 is the ideal choice for the companies.

From the micro perspective, Industry 4.0 represents integration of smart industries both horizontally and vertically. The latter covers various value creation units at a lower aggregation level, such as manufacturing blocks, manufacturing lines, and manufacturing stations (Stock & Seliger, 2016). Since smart factories will be using renewable energies to maintain their self-sufficient supply, they will be both suppliers and consumers to meet energy requirements (Berger, 2014). The horizontal integration within smart factories, from the micro perspective, will be possible through the intersection of value creation modules across the material inflow and outflow (Stock & Seliger, 2016). The inbound and outbound flow of goods is carried out by transport equipment, regulated by automated transportations, such as Automated Guided Vehicles (AGVs) to practicalize decentralized coordination of supplies and products (Stock & Seliger, 2016). Vertical integration, on the other hand, implies the intelligent networking of the value creation factors like product, equipment, and human, besides the different aggregation levels of the value creation modules, like marketing and sales, service, procurement, from manufacturing stations up to the smart factory (Porter, 2011).

The literature review shows that many studies have been carried out on industry 4.0 and the earlier revolutions. Still, there are not many studies that summarize the prospects and risks of industry 4.0. The present study aims to fill this research gap and contribute to the fourth industrial revolution's existing knowledge. The literature review establishes the potential of the fourth industrial revolution that reaps the emergent and disruptive technologies to accomplish ever-higher production efficiencies. It assesses how the industry's capabilities accommodate and adapt to society's changing needs by digitalizing the economy permanently. It researches the characteristics and application aspects of the flourishing of industry 4.0 by investigating the prospects and challenges regarding the implementation of the specific strategies in facilitating the socioeconomic development of nations. It navigates the possibilities, opportunities, trends, and threats of business transformation and economy in the context of 4.0 industrialization by identifying the digital interconnection across the products, business models, and value chain. It sets a new direction toward a hyper-connected society and economy. Based on the studies' implications and conclusions, the literature review demonstrates how industry 4.0 can adjust to volatile, dynamic market environments and inter-industrial networks because of the mechanics of transformability and flexibility. It also calls for addressing the risks and obstacles of implementing industry 4.0 at the strategic level, due to lack of technological infrastructure.

3   RESEARCH METHODOLOGY

The objective of this study is to determine what has been researched so far about industry 4.0. Industry 4.0 is in the nascent stage of implementation. A broad review of periodical articles, seminar papers, books, and edited volumes was done to accomplish the goal. As the topic is very innovative and significant, so a comprehensive literature review on Industry 4.0 was conducted by employing search engines: EBSCO host, ProQuest, Science of Web, Scopus, and Google Scholar. Due to the advances in technology, a tremendous amount of data is available, which can be used for research (Andrews et al., 2012; Smith, 2008; Smith et al., 2011). Also, relevant reports on Industry 4.0 from influential consultancy institutions, such as Boston Consulting Group, Deloitte, and Mckinsey, were investigated by Governmental and EU bodies. Secondary analysis is also a systematic process with procedural and evaluative steps (Johnston, 2014). The notable advantage of secondary data is the cost-effectiveness and accessibility (Dale et al., 1988; Glaser, 1963; Smith, 2008). Secondary data investigation offers a methodological benefit and can also contribute to research by fortifying its knowledge resource. In addition to systematic reviews, other methodologies used were quantitative electronic survey and primary qualitative interviews, and qualitative case studies with a traditional qualitative approach. The survey questionnaires were circulated via social media platforms and were returned to knowledge management administrators and experts. The survey was conducted using the Six Sigma Quality Initiative Method principles and a data-driven process called Define, Measure, Analyze, Improve, and Control (DMAIC). The methodological framework's empirical focus was to investigate the classification of the critical indicators of testing technological, environmental, and socioeconomic consequences resulting from the precipitation of the fourth industrial revolution. The survey questions were related to the strategic opportunities of industry 4.0 and the implementation prospects; the operational opportunities of industry and their implementation potential; the environmental, economic, and social opportunities or consequences of industry 4.0; the industry's competitiveness and viability; and employee readiness to accept it and the required qualifications to integrate it. Content analysis to evaluate several case studies and systems analysis of statistical indicators were some other methods employed to gage the readiness to adapt to fourth industrialization (Rumi et al., 2020; Gladkikh, 2020). The study also examined the latest edition of the Global Information Technology Report (GITR) published by the World Economic Forum jointly with ISEAD—the Business School for the World to study the preparedness for the fourth industrial revolution among Gulf Cooperation Countries (GCC) and Brazil, Russia, India, China, and South African (BRICS) countries (Baller et al., 2016).

4   RESULTS

The e-survey, through its demographic questions, helped us assess the business compass of the 4.0 industry, the dimensions of teams and organizations, and budget estimates. It gave insights to compare the fourth industry's prospect of business with other business sectors, like legal, financial, and healthcare. It allowed us to craft value creation by evaluating the feedback from responders. Based on the data of interviews, e-survey, and literature review, we conclude the industry 4.0 will lead to increased income inequality. The studies of Kuzmenko and Roienko (2017) have suggested in their forecasting results, after adjusting the Gini ratio with the Fourth Industrial revolution, that the new technologies will overtake and surpass high-skill workers. It will also replace low-skilled laborers. Income inequality is likely to intensify in industrialized counties like Germany. The primary interviewees agreed on developing communication, collaboration, and flexibility skills besides technical skills to fit into the new era of fourth industrialization (Carter, 2017). The interviewees also predicted and acknowledged the centrality of emerging trends in the next 1–2 years, such as connected and linked data, more innovation and technical improvements, vendor competition, mobile fast, and big data (Carter, 2017). However, very few interviewees did report about the need for membership of information professional organizations. A qualitative investigation of study findings reveals that despite industry 4.0 holds immense opportunities in sustainable Information and Communication Technologies (ICT) development, increase in productivity, women, and citizen empowerment, yet it might contribute to social inequality, colossal unemployment, a threat to social security, the domination of MNCs, and the ruining of the economy (Rumi et al., 2020). The DMAIC results gave a roadmap to understand the industry direction and identify challenges. It highlighted the significant challenges in implementation, such as information technology security issues, privacy and data protection problems (Alaloul et al., 2020).

Based on the secondary data of the literature review findings on Industry 4.0, its prospect in GCC is less promising unless the government intervenes actively and elaborates on industrial and economic policies to achieve a dynamic ecosystem and interactive environment for smooth flow of data across the production chain (Poma et al., 2020). Similar is the case with Italy, which, although structurally different from UAE, faces systemic and organizational bottlenecks and requires government involvement to appropriately utilize significant data architecture (Poma et al., 2020). The GITR analysis finds that the GCC countries are doing better than BRICS to welcome 4.0 industrialization. The former has better political and business environments, better infrastructures, reasonable affordability, needed skills, usage of ICT by the individual, business sectors, and government sectors than the latter (Alshubiri et al., 2019). Qatar has done exceptionally well to integrate the new industry because of excellent environmental factors, the country's political and regulatory environment, business, innovation, the availability of the latest technology, education, and business procedures (Kaba & Said, 2020).

4.1   The Prospects of Industry 4.0

According to Gershenfeld and Vasseur (2014), 3D printing for prototyping will reduce inventors and markets' challenges. The author also states that the latest technologies will benefit tissue scientists who can use 3D printing techniques for porous scaffolds. The latest know-how will act as enablers for entrepreneurs to start small companies with low startup costs. The entrepreneur can create his product with 3D printing without time challenges and outdated production methods. According to Korea health industry development institutions (2016), artificial intelligence and big data can transform the connectivity between healthcare and the sick. The development of intelligent sensors in the 2020s will connect homes and healthcare providers with data-based services. As stated by the organization for economic cooperation and development, artificial Intelligence will help combat hospital-centered medical care in the technology-driven age. Schwab (2015) quotes that the fusion of innovative technologies will integrate various scientific and technical domains. This fusion will have a synergetic effect. It will be more than a complementary technology as it will create new untapped markets and growth opportunities for all the participants in the innovation games. Technology fusion will create incremental benefits from previously separated fields to create a product.

Based on Ben and Mondada (2018) studies, Robotics will have a dramatic influence on our life. Robots are automated motorized tools. Robots are used in many places like educational institutions, households, healthcare, and outer space. They can cook, play music, and drive cars, follow human instructions, assist doctors in surgeries, improve the quality of our lives at home, work, and play. Nowadays, robots are used in arial photography pipe inspection and bomb disposal, which rely on operators' devices. Most of the robots are not autonomous but perform sub-tasks automatically. For example, the autopilot of a drone stabilizes the flight while the pilot chooses the flight path. Robots are used to clean swimming pools and lawnmowers. Some robots have advanced sensing feature, which is used to identify and remove weeds from agricultural land.

Access to information through automotive computer devices, digital facilities, and mobile phones can provide massive educational opportunities to people living in underdeveloped countries. Through social media like Facebook Twitter, people can interconnect instantly to any part of the world and share their views and suggestions. Most people across the globe can access world events through social media. People have access to new products and markets, and businesses can flourish using the digital marketing platform. There is an opportunity to learn and earn more using new ways and improve their personalities, which was impossible in earlier times. (Schwab, 2015) Finally, the fourth industrial revolution will completely transform the human civilization. It will impact all spheres of our existence such as our privacy, ownership, work and life balance, career and skill development along with interpersonal relationship.

New technologies have influenced the marketing and delivery of goods to customers. Drones are used as delivery vehicles these days, which can distribute products to remote areas, also in less time and improving the economies of rural and small sizes. Self-directed vehicles can redesign the existing spaces of the congested cities and make more space available for more communal and human-centered areas. The road mortalities, carbon emissions, and insurance costs can be decreased by the developments in locomotive security provided by the 4.0 industrial revolution. Digital technology can do the automatable jobs at a faster speed. It can release the workforce to focus and address more intricate business matters and give them more independence. Workers can concentrate on more creative resolutions for the earlier insoluble problem.

The fourth IR is a unique opportunity provider. Though most of society's stakeholders view it as a challenge, it offers a great opportunity. It presents an opportunity to reconsider how skills are realized in jobs and how they can create pathways for an inclusive approach toward marginalized youth globally. The business community should seize 4IR opportunities and gain employment and promote individual economic mobility. 4 IR will generate employment opportunities and has transformed focus on jobs that call for technical skills. Interconnectivity and innovations have created more scope for entrepreneurship. It will provide a unique opportunity for the young generation across the globe to bypass the traditional barriers of entry to employment. Entrepreneurs will be able to leverage social knowledge and resources. Will create new opportunities for work-based learning and industry-driven demand analysis.

4.2   Risks Associated with Industry 4.0

4.2.1   Changes in Employment and Human Values

Artificial intelligence is expanding and enhancing human lives in many ways, but excessive use of technology can negatively affect human lives, as noticed in previous industrial revolutions. Industry 4.0 can disrupt people from jobs and create questions on the relationships of human beings and machines. Over dependency on devices can drive away from the human qualities of adjustment, emotional balance, openness, and agreeableness. Different types of tasks done by artificial intelligence and automation can affect the job market. Less educated people will have the disadvantage of learning a new method of doing work. Training people for the changing nature of work will be a challenge for organizations and governments in terms of cost and time. The accessibility of training institutions for such advanced and latest skill sets will not be accessible in terms of price and availability to a more extensive section of the population.

4.2.2   Change in Confidentiality

In today's technological world, providing and delivering more personalized services is crucial to trace a person's private information. For example, social media like Facebook trails what an individual is browsing to recommend content and advertisement that may be of interest to that person. Before going to the shop, the salesman will know about the customer's marital status, purchase history, and choices about the products. To increase sales, the vendors investigate your past purchasing habits and endorse products or services, which are more suitable for the customer. Human beings' activities in the future can be monitored by the billions of 3D printed "smart dust cameras", which can be useful to keep us safer by informing us about natural disasters, tracking criminals, and giving traffic reports. Still, each human being's activity can be on surveillance when they do not want to be observed.

4.2.3   Changes in Faith

"The biggest global issue is continued erosion and trust". According to Claus, Schwab Trust is the factor that combines societies and brings stability in the world. In today's world businesses, govt. media and even technology are losing public trust. The technologies used in the 4th industrial revolution are unbiased in themselves, but their applicability shapes faith among people. General business confidence, the government, the media, and even technology are falling. The crisis that is dividing societies and creating instability around the world. The question arises, whether robotic systems, artificial intelligence will make people's lives better, or are they going to control human lives and make them afraid of machines? Or how people are going to trust the institutions and service providers who assemble and preserve their data. Governing and managing the technologies of the 4th industrial revolution will be a big challenge, and these technologies will dominate those who regulate them.

4.2.4   Changes in Equality

Innovation is increasingly facilitated using technological development and world is becoming smaller and smaller. But global society do not need to become more inclusive and more associated. The previous industrial revolutions have concluded that technology has concentrated wealth generation to small, powerful groups above the rest. Powerful technological changes and global digital networks make us susceptible to corporal and cyber-attacks and keep humans under unnecessary surveillance. According to the world economic forum, the 4th industrial revolution can improve the quality of living and improve people's income levels through unexpected innovations and technology's power. Still, it is concentrating the income in the hands of a small group of people, increasing the inequality and leading to political divergence, social disintegration, and absence of confidence among institutions. Previous revolutions have not benefited all the people in the world at the same pace. (Schwab, 2016) The authors of "Shaping the Fourth Industrial Revolution" pointed out that out of 2.4 billion people, one-third of the total population, do not have access to safe sanitation and clean drinking water, and 1.2 billion people do not have electricity inventions of the second industrial revolution. At least 600 million people are untouched even by the mechanization of the first industrial revolution. So, it should be ensured that the benefits and opportunities of industry 4.0 should be equally distributed among the people and societies who are still lacking the benefits of the first, second, and third revolutions.

5   Conclusion

The paper mainly focuses on the 4th industrial revolution called Industry 4.0, which allows connectivity across systems, automated manufacturing, and customized productivity at a reasonable cost. This paper discusses the prospects and risks of the 4th industrial revolution. The article makes conclusions based on secondary data available in the research papers, EBSCO, and other online platforms, journals, etc. It also uses an electronic questionnaire, survey, qualitative interviews, content analysis, and technology reports to inquire into the industry's scope, opportunity, and threat. The fourth industrial revolution describes how technology will transform what we do at home and work, and change our lives dramatically. The new technologies of industry 4.0 can be life-changing agents for humanity both positively and negatively. It can enhance the different aspects of human life, and provide massive opportunities in education, health, manufacturing, sciences and, also enable the benefits of connectivity to reach people living in underdeveloped countries. But it can have negative impacts like the previous industrial revolutions. Industry 4.0 can disrupt people from jobs and raise questions on the connection between human beings and machines. Over-dependence on devices can drive away human qualities of adjustment, emotional balance, openness, and getting along with other people. The prospects and risks in the research paper are from a macro perspective. The purpose of the study is to examine the prevailing literature and express views thereof. Still, methods like surveys have not been utilized for this study. Additional investigations on the implications of industry 4.0 in diverse countries with the varied geopolitical environment and different sectors with numerous economic parameters is needed.

References

Alaloul, W. S., Liew, M. S., Zawawi, N. A. W. A., & Kennedy, I. B. (2020). Industrial revolution 4.0 in the construction industry: Challenges and opportunities for stakeholders. *Ain Shams Engineering Journal, 11*(1), 225–230.

Alshubiri, F., Ahsan Jamil, S., & Elheddad, M. (2019). The impact of ICT on financial development: Empirical evidence from the Gulf Cooperation Council countries. *International Journal of Engineering Business Management, 11*, 1–14. https://doi.org/10.1177/1847979019870670.

Andrews, L., Higgins, A., Andrews, M. W., & Lalor, J. G. (2012). Classic grounded theory to analyse secondary data: Reality and reflections. *Grounded Theory Review, 11*(1).

Baller, S., Dutta, S., & Lanvin, B. (2016). *The global information technology report 2016: Innovating in the digital economy.* World Economic Forum. https://doi.org/10.1111/j.1432-1033.1993.tb17792.x.

Ben-Ari, M., & Mondada, F. (2018). *Elements of Robotics.* https://doi.org/10. 1007/978-3-319-62533-1_1.

Berger, R. (2014). INDUSTRY 4.0–The new industrial revolution: Maschinenbau| Engineered Products/High Tech| Branchenexpertise| Expertise. Roland Berger.

Butler-Adam, J. (2018). The Fourth Industrial Revolution and education. *South African Journal of Science, 114*(5/6), Art. #a0271, 1. https://doi.org/10. 17159/sajs.2018/a0271. Retrieved October 30, 2018.

Carter, D. (2017). Creativity in action–the information professional is poised to exploit the fourth industrial revolution: The business information survey 2017. *Business Information Review, 34*(3), 122–137.

Dale, A., Arbor, S., & Procter, M. (1988). *Doing secondary analysis.* Unwin Hyman.

Feinstein, C. (1998). Pessimism perpetuated: Real wages and the standard of living in britain during and after the industrial revolution. *Journal of Economic History, 3,* 58.

Fonseca, L., Ramos, A., Rosa, A., Braga, A. C., & Sampaio, P. (2016). Stakeholders satisfaction and sustainable success. *International Journal of Industrial and Systems Engineering, 2*(2), 144–157.

Glaser, B. G. (1963). Retreading research materials: The use of secondary analysis by the independent researcher. *The American Behavioural Scientist, 6*(10), 11–14.

Gershenfeld, N., &Vasseur, J. P. (2014). *As objects go online: The promise (and pitfalls) of the Internet of Things.* Retrieved from https://www.foreignaffairs. com/articles/2014-02-12/objects-go-online.

Gladkikh, E. G. (2020). Prospects for the development of the Volgograd region within of the concept of the Industrial Revolution 4.0. *Competitive Russia: Foresight Model of Economic and Legal Development in the Digital Age,* 86.

Hessman, T. (2013). The down of a smart factory. *Industry Week.* Retrieved from http://www.industryweek.com/technology/dawn-smart-factory.

Johnston, M. P. (2014). *Secondary data analysis: A method of which the time has come.* University of Alabama.

Kaba, A., & Said, R. (2020). *Assessing readiness for the Fourth Industrial Revolution: A comparison of GCC and BRICS countries.*

Kagermann, H. (2015). Change through digitization—Value creation in the age of Industry 4.0. In H. Albach, H. Meffert, A. Pinkwart, & R. Reichwald (Eds.), *Management of permanent change* (pp. 23–45). Springer.

Kagermann, W. W. H., & Helbig, J. (2013). *Recommendations for implementing the strategic initiative Industrie 4.0.* Available from http://www.acatech.de/ fileadmin/user_upload/Baumstruktur_nach_Website/Acatech/root/de/Mat erial_fuer_Sonderseiten/Industrie_4.0/Final_report_Industrie_4.0_accessible. pdf.

Kane, G. C., Palmer, D., Phillips, A. N., & Kiron, D. (2015). Is your business ready for a digital future? *MIT Sloan Management Review, 56,* 37.

Korea Health Industry Development Institute. (2016). *Changes in the fourth industrial revolution and health industry paradigm.* Korea Health Industry Development Institute.

Kuzmenko, O., & Roienko, V. (2017). Nowcasting income inequality in the context of the Fourth Industrial Revolution. *SocioEconomic Challenges, 1*(1), 5–12.

Lasi, H., Fettke, P., Kemper, H.-G., Feld, T., & Hoffmann, M. (2014). Industry 4.0. business & information systems engineering: *The International Journal of WIRTSCHAFTSINFORMATIK, 6*(4), 239–242. https://doi.org/10.1007/s12599-014-0334-4.

Mokyr, J. (1985). *The new economic history and the Industrial Revolution.* Rowan & Littlefield Publishers Inc.

Mosconi, F. (2015). *The new European industrial policy: Global competitiveness and the manufacturing renaissance.* Routledge.

Organization for Economic Co-operation and Development. (2016). *Health reform: Meeting the challenge of ageing and multiple morbidities.* Organization for Economic Co-operation and Development.

Poma, L., Shawwa, H. A., & Maini, E. (2020). Industry 4.0 and big data: Role of government in the advancement of enterprises in Italy and UAE. *International Journal of Business Performance Management, 21*(3), 261–289.

Porter, M. E. (2011). *Competitive advantage of nations: Creating and sustaining superior performance.* Simon and Schuster.

Posada, J., Toro, C., Barandiaran, I., Oyarzun, D., Stricker, D., de Amicis, R., & Vallarino, I. (2015). Visual computing as a key enabling technology for industrie 4.0 and industrial internet. *IEEE Computer Graphics and Applications, 35*(2), 26–40.

Rumi, M. H., Rashid, M. H., Makhdum, N., & Nahid, N. U. (2020). *Fourth Industrial Revolution in Bangladesh: Prospects and challenges.*

Schlechtendahl, J., Keinert, M., Kretschmer, F., Lechler, A., & Verl, A. (2015). Making existing production systems Industry 4.0-ready. *Production Engineering, 9,* 143–148. https://doi.org/10.1007/s11740-014-0586-3.

Schlick, J., Stephan, P., & Zuhkle, D. (2012, August). *Produktion 2020. Auf dem Wegzur 4.0. industriellen Revolution. IM—Fachzeitschrift fur Information Management und Consulting.*

Schwab, K. (2015). *The Fourth Industrial Revolution: What it means and how to respond.* Retrieved from https://www.foreignaffairs.com/articles/2015-12-12/fourth-industrial-revolution.

Schwab, K. (2016). *The Fourth Industrial revolution.* World Economic Forum, 9193 route de la capite.

Schwab, K. (May 25, 2018). The Fourth Industrial Revolution. *Encyclopaedia Britannica*. Available https://www.britannica.com/topic/The-Fourth-Industrial-Revolution-2119734. Retrieved September 1, 2018.

Smith, E. (2008). *Using secondary data in educational and social research*. McGraw-Hill Education.

Smith, A. K., Ayanian, J. Z., Covinsky, K. E., Landon, B. E., McCarthy, E. P., Wee, C. C., & Steinman, M. A. (2011). Conducting high-value secondary dataset analysis: An introductory guide and resources. *Journal of General Internal Medicine, 28*(8), 920–929. https://doi.org/10.1007/s11606-010-1621-5.

Sommer, L. (2015). Industrial revolution—Industry 4.0: Are German manufacturing SMEs the first victims of this revolution? *Journal of Industrial Engineering and Management, 8*, 1512–1532. https://doi.org/10.3926/jiem.1470.

Stock, T., & Seliger, G. (2016). Opportunities of sustainable manufacturing in industry 4.0. *Procedia Cirp, 40*, 536–541.

Vogel-Heuser, B., & Hess, D. (2016). Guest editorial Industry 4.0–prerequisites and visions. *IEEE Transactions on Automation Science and Engineering, 13*(2), 411–413.

World Economic Forum. (2016). *Futures of work: Lifelong learning is the new Black*. Available https://www.the4thindustrialrevolution.org/the-world-economic-forums-8-futures-of-work-lifelong-learning-is-the-new-black/. Retrieved October 15, 2018.

Yu, J., Subramanian, N., Ning, K., & Edwards, D. (2015). Product delivery service provider selection and customer satisfaction in the era of Internet of things: A Chinese e-retailers' perspective. *International Journal of Production Economics, 159*, 104–116. https://doi.org/10.1016/j.ijpe.2014.09.031.

Employment and Human Resources

Re-Inventing Human Resource Management Through Artificial Intelligence

Rabia Imran

1 Introduction

The competitive business world demands the organizations to transform their processes according to the changing demands. The increase in technology has made the world smaller. The organizations nowadays are no more threatened due to the local competitors, but the competition has also grown across the borders. As a result, enhancing technological level to maintain a competitive advantage has become a key for development. Globalization coupled with dynamics of the knowledge economy has increased the need of digital technologies. These Digital technologies are now changing everything starting from how people live, communicate with each other to how the businesses are conducted. Digitalization is taking place everywhere. In this regards a concept emerged and suddenly grabbed the attention of academicians and practitioners is termed as artificial intelligence.

Human resources are considered as one of the precious assets of any organization. It is an important duty of the organization to take care of

R. Imran (✉)
Department of Management, College of Commerce and Business Administration, Dhofar University, Salalah, Sultanate of Oman
e-mail: rimran@du.edu.om

this asset. In this regard, organizations perform a broad range of tasks such as staffing (recruitment and selection), performance and compensation management and employee development. To fulfill these tasks, the human resource management relies on different techniques, sometimes borrowed from other disciplines such as psychology, sociology and operations research. However, nowadays the techniques borrowed from information systems like artificial intelligence (AI) are playing important role (Jantan et al., 2010). Artificial intelligence (AI) involves machine assistance or sometimes replacement in the tasks solely done by humans earlier. This brings about major changes in the task performance especially in human resource management. AI offers many techniques for specific tasks involved in HRM such as recruitment, selection, training, and development (Chien & Chen, 2008; Giotopoulos et al., 2005; Kaczmarek et al., 2010).

The aim of this chapter is to explore the change brought in Human resource management due to adoption of AI techniques. The chapter is also aimed at identifying the current scenario, the fit between HR tasks and AI, benefits and challenges of AI, and its future role in HRM.

2   ARTIFICIAL INTELLIGENCE

Artificial intelligence (AI) is a concept that was present for a long time however, recently, the world has started realizing its importance (Tecuci, 2012). Artificial intelligence is a branch of computer science which has quickly grabbed the business world and whose sole aim is to create intelligent machines. It is the term used for machines and computer related interventions and was first introduced in an academic conference by John Mccarthy in 1956 though the exploration of this concept was initiated much earlier (Verma & Bandi, 2019). It is not only the essential part of technology industry, but it has started becoming a necessity for other industries to ensure their sustainable performance (Habeeb, 2017).

There is no one particular definition of artificial intelligence. The concept is a combination of two words artificial and intelligence. The world artificial is something that has not occurred naturally rather it is human made or produced. The copy of things that occurred or produced naturally also come into this category (Oxford Dictionary, 2019b). Artificial means having human-made interventions to trigger naturally occurring phenomenon (Johansson & Herranen, 2019). The main problem comes while defining the term intelligence. The literal meaning of the word is the ability to do well. It is the capability that helps to learn, think, and then understand in a logical way (Oxford Dictionary, 2019a). However, in the context of artificial intelligence this term is used to identify intelligence in relation with artificial things. Intelligence here can be defined as the ability to think, do planning, understand the data and generalize in timely manner with the available information and in a given situation (Kaplan, 2016; Legg & Hutter, 2007; Ved et al., 2016). Artificial Intelligence on the other hand is a combination of these two concepts and is described as the ability of human-made things such as machines to discover, infer, and comprehend by themselves in a way that is similar to humans (Johansson & Herranen, 2019).

2.1   Branches of Artificial Intelligence

McCarthy (2004) has identified branches of Artificial intelligence. Some of them have grown up to full branches. However, some are still regarded as important aspects. They are listed as follows.

2.1.1   Logical Artificial Intelligence
It is a branch of AI that has a mathematical language. It is an idea that characterizes knowledge of an agent about its world. It logically presents its goals and existing situation and takes a decision of what needs to be done for goal attainment on the basis of the logic (McCarthy, 2000; Minker, 2012).

2.1.2   Search
It is an important concept in problem solving aspect of artificial intelligence. Large numbers of possibilities are examined by artificial intelligence programs and continuous breakthroughs are made for its more effective use in a variety of domains (Kanal & Kumar, 2012).

2.1.3   Pattern Recognition
Numerous insights are yielded due to pattern recognition in the fields of human visual sensory and perceptual processes. It is an effective method of automatic knowledge acquisition (Chen, 2013). A program, while making comparisons, can tend to contrast different concepts with a pattern, e.g., face recognition tries to match the patterns of eyes and nose.

2.1.4   Representation
This branch of artificial intelligence is used for representing the knowledge and facts in an appropriate way. Mathematical logic language is generally used for this branch.

2.1.5   Inference
Some things are facts whereas, others can be inferred. This branch provides a set of rules to deduce from existing knowledge base to infer new information. It is one of the most important reasoning methods (Aronson et al., 2005).

2.1.6   Common Sense Knowledge and Reasoning
This branch involves the knowledge about the facts and capability of humans to use it and make presumptions about the daily life situations. Real world knowledge and reasoning abilities are the base for performing various intelligence tasks like understanding the texts, planning activities, and computer vision. Intelligence machines need to better understand human commonsense to perform such tasks. The techniques used for such tasks are web mining, logical analysis, crowd sourcing, etc. (Davis & Marcus, 2015).

2.1.7   Learning from Experience
This is the basic feature. Programs have limited learning abilities and those are dependent on the facts or behaviors represented by their formalism. If there is a repeated performance of the task, then next time the outcome is slightly improved by modifying the system.

2.1.8 Planning

It is an important element in the artificial intelligence. It starts with adequate situation-specific knowledge that results in a pre-arrangement of sequence of actions for goal achievement.

2.1.9 Epistemology

This concept emerges from psychology and deals with knowledge theories. Artificial intelligence gains its strength through the initial understanding of epistemology especially the knowledge specific to problem solving.

2.1.10 Ontology

The concept started gaining attention in 1990s. It deals with formally representing knowledge, developing concepts, and building associations between them. In artificial intelligence this concept is used for dealing with the properties and specifications of a variety of objects.

2.1.11 Heuristics

This is a unique way of problem solving once traditional ways are either not working or are too slow. It is an essential part of artificial intelligence. It is discovering the ideas set in the programs and produce solutions in a reasonable amount of time.

2.1.12 Genetic Programming

It is a revolutionary technique in problem solving and has a broad range of application domain. It is a technique of evolving programs and is fit for tasks that need solutions.

Table 1 reveals the summary of few Proven Artificial Intelligence Techniques.

2.2 Techniques of Artificial Intelligence

Table 1 Summary of few Proven Artificial Intelligence Techniques

Proven AI Techniques	
Representation	Languages, Domain Modeling, and Knowledge Engineering Rules, frames, classes, cases, hierarchies, propositions, constraints, demons, certainty factors, fuzzy variables
Inference	Theorem-Proving, Heuristic Reasoning, and Matching Techniques Forward and backward-chaining, unification, resolution, inheritance, hypothetical reasoning, constraint propagation, case-based reasoning
Control	Goal and data directed, messaging, demons, focus, agenda, triggers, meta-plans, scheduling, search algorithms
Problem Solving Architectures	Rule based, object oriented, frame based, constraint based, blackboard, heuristic classification, task-specific shells

Source Hayes-Roth (1997, p. 104)

3   HUMAN RESOURCE MANAGEMENT

Human resource management is considered as a backbone of any organization. Changing dynamics of the work environment now emphasize more on having the right person on the right job. This requires the human resource practices to be relevant to the changing needs. The concept of human resource management, despite being widely researched, is still an interesting one and has a lot of room to explore. It is defined as the ability to gain and develop talented workforce that can execute goals and strategies of an organization. It can be considered as an organization's effort for acquisition and maintenance of new abilities, skills, and competencies in the workforce through various management practices (Johansson & Herranen, 2019).

A range of practices constitute the process of human resource management. However, some popular ones are recruitment, selection, training, development, and compensation (Wall & Wood, 2005). The focus of these practices is on employee retention at a satisfactory level. HRM role in organizations has changed drastically during past few years. It is becoming an issue of strategic importance for the organizations as in the knowledge world competent employees are the one that can guarantee competitive advantage and sustainable performance. The most important practices in these regards are recruitment, selection, training, and development as these are the foundation of any other HR practice. Without the right pool of applicants as well as proper selection of techniques and company would be unable to hire the right person for the right job thus all other practices would be ineffective. With the recent focus on employee development the focus on training and development has become immense. The organizations are now considering training and development as the most important HR function. The organizations adopt various HR practices to enhance their performance. A few very important HR practices are discussed below.

3.1   Important HR Practices

3.1.1   Recruitment

Last few decades are evident of the increased attention of the researchers and practitioners on the concept of recruitment. This is the initial and one of the most important practices that effects the behaviors within an organization (Taylor & Collins, 2000). The care taken in attracting a right pool of applicants depicts the quality of workforce hired. Recruitment is described as efforts exerted in attracting relevant candidate pool for any available opportunity (Stoilkovska et al., 2015). It holds a central position in HR practices as the quality of employees hired determines the effectiveness of other practices (Griepentrog et al., 2012). The following Fig. 1 shows the recruitment process that is available in existing literature and can be used for recruitment in any domain.

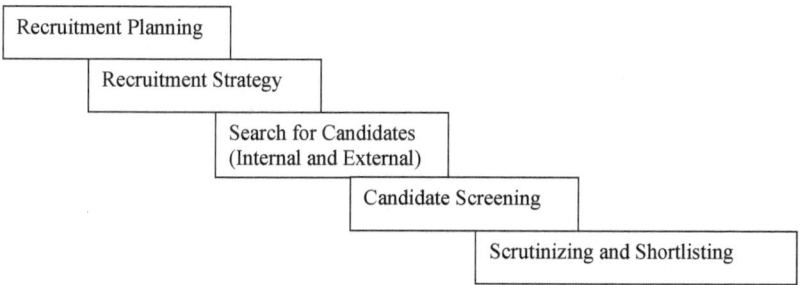

Fig. 1   Recruitment process (*Source* Compiled by the author)

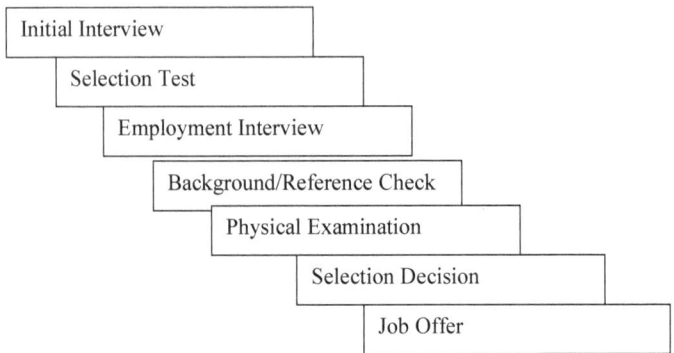

Fig. 2 Selection process (*Source* Compiled by the author)

3.1.2   Selection

Selection takes the second position in HRM practices as the pool of applicants need proper techniques to choose the best out of the whole lot. After the pool is built then the organization deploys appropriate techniques to select the right person for the available vacancy (Newell, 2005; Stoilkovska et al., 2015). The techniques used for selection may include one or a combination of written test, interviews, management games, simulations, etc. However, the organization must ensure equal chances of selection for all the candidates. A typical selection process is shown in Fig. 2.

3.1.3   Training and Development

Due to the strategic importance of learning organizations training and development is considered as value adding HR practice (Berk & Kaše, 2010). It is defined as a planned effort to enhance learning required for job (Noe & Kodwani, 2018). It is not only used to develop job specific skills but also to enhance already available set of skills. This is one of the most important HR practices that are needed to develop human capital that leads to desired organizational outcomes (Sanders & Murphy-Brennan, 2010). A typical training process adopted by the HR department is described in Fig. 3 as follows.

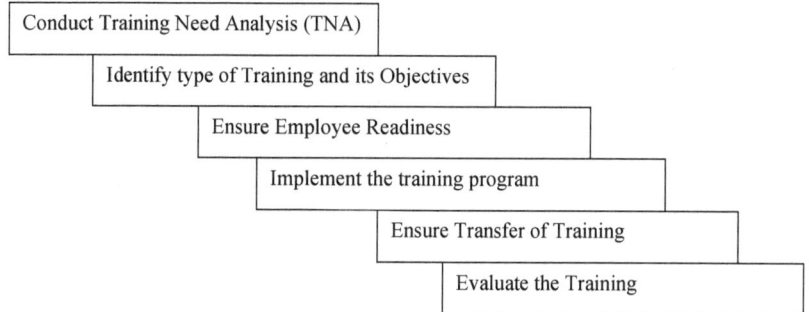

Fig. 3 Training and development Process (*Source* Compiled by the author)

4   Impact of Artificial Intelligence on HRM

With increasing competition, the organizations struggle for achieving competitive advantage to reach sustainable performance. Thus, the focus on acquiring and retaining talented human resources within an organization has increased. This has led to the adoption of unconventional and up-to-date HR practices (Laha, 2019). The major challenge in the current environment is to bring computers and humans together for developing a competitive advantage.

4.1   The Fit (AI and HRM)

In order to use any technology for specific tasks a common base is needed. For this purpose, the task technology fit approach becomes a useful foundation (Goodhue & Thompson, 1995). In simple words this approach means the connection between the requirement of the task and technology functionality (Goodhue & Thompson, 1995). This approach is common and has been used in different domains which make it important to examine the potential effect of AI techniques on human resource management (Furneaux, 2012). Human resource management is generally described as subset of tasks also called practices supporting organizational strategy and performance. These tasks or practices can be supported by technology either using information, automation, or machine replacement (Zuboff, 1985).

4.2   The Current Scenario (AI and HRM)

Artificial Intelligence is now being considered as an important factor in transforming and supporting HRM functions such as recruitment and selection, training and development, employee retention and talent management, etc. (Merlin & Jayam, 2018). All sorts of interventions that are named as artificial intelligence now are machine learning and the employers are transforming work experience by using AI (Sierra-Cedar, 2018–2019). The firms adopting AI explore its application in HRM. This replaces certain functions of human resource management and, thus, making them more efficient. It has transformed the way organizations work. All the organizational functions have started operating better because of artificial intelligence.

The major effect of artificial intelligence has been on the recruitment process. Since long time the HR processes were evolving, and organizations were switching to online recruitments. However, with the advent of artificial intelligence recruitment professionals took a sigh of relief. The extraction technique made resume and relevant information scanning an easy task. The increased number of applicants for many jobs makes this task hefty and troublesome. However, switching from manual system to the choices provided through AI makes this task efficient and effective. Another benefit of AI driven recruitments system is that candidate ranking system which helps in rank order the applicants by giving them scores. Chat boats are another AI driven recruitment assistant that improves connectivity with the candidates through text messages and emails. Job matchmaking techniques also introduced that make resume' sorting an easy job. The selection process can also be efficient with the use of video job interviews that can increase the pool base and can make available global talent for the efficient working of the organization. The AI interventions have been successfully used in training and development processes. The organizations are using the online trainings and developmental tools for better performance of their organization.

The following Table 2 reveals some AI based interventions in HRM.

Table 2 AI Interventions in HRM

Execution technique	Automated scanning of relevant information (Faliagka et al., 2012)
Ranking system	Rank order the applicants (Faliagka et al., 2012)
Chatboats	Improved connectivity (Upadhyay & Khandelwal, 2018)
Job Matchmaking techniques	Resume sorting (Montuschi et al., 2014)
Video job interviews	Job interviews via video conferencing (HireVue, 2018)

Source Compiled by the author

4.3   The Benefits of AI Based HRM

AI based HR practices bring a lot of benefits for organizations. These interventions make the work easy for the HR team and save expenditure by reducing the paperwork. The access to the information, scanning it, and sorting it becomes easy. Prompt feedbacks of queries can be generated, and a lot of time can be saved. The following Table 3 reveals the benefits of AI based HRM.

4.4   The Challenges of Implementing AI in HRM

Everything that exists in the world has a positive and a negative side. Artificial intelligence-based intervention is no exception. There are some challenges that are faced while this system is put into practice. The AI based recruitment and selection system despite have numerous benefits

Table 3 Benefits of AI based HRM

Easy work	Dickson and Nusair
Less paperwork	(2010)
Save expenditure	
Easy access to information	Upadhyay and Khandelwal
Prompt feedback to queries	(2018)
Effortless scanning of information	Faliagka et al. (2012)
Time saving	Guchait et al. (2013)
Improved performance	Martincevic and Kozina (2018)

Source Compiled by the author

posit some challenges as well. It can expose to the threat of compromising personal privacy. Private information like identity card and social security numbers, resident details, etc., can be leaked. Moreover, there is a chance of unconscious biasness due to restrictive settings in the system. The rapport building and negotiations also have no room in the automated system. The following Table 4 reveals the challenges of AI based selections and recruitment process.

The HR practices and systems that have AI interventions may have some inbuilt biases. It can bypass some unique but relevant information. For example, the candidate will fill only the information that is pre-determined. In case he/she wants to convey some extra but relevant information they would not be able to do it due to the pre-determined restrictions. There is a high chance of stereotyping with these kinds of systems. Minority groups of any time would be in a disadvantage. Following Table 5 reveals the biases of AI based HRM.

4.5   The Re-Invention of HRM (The Future)

AI has its future in Human Resource Management. It can re-invent the wheel with a lot of updating and advancements. AI can enhance almost all functional areas of HRM (HR.com, 2017). In future, with the help of AI HRM may be transformed and have an HR world free of biases. Human nature is prone toward biases; however, there is no room for

Table 4 Challenges of AI based HRM

Compromising of personal privacy	Bondarouk and Brewster (2016)
Unconscious Discrimination	Stuart and Norvig (2016)
No room for negotiation and rapport building	Upadhyay and Khandelwal (2018)

Source Compiled by the author

Table 5 Biases of AI based HRM

Bypass of unique but relevant information	Upadhyay and Khandelwal (2018)
Unconscious Biasness and stereotyping	Beattie and Johnson (2012)
Exclusion of minority groups	

Source Compiled by the author

biases with artificial intelligence. The machine-based interventions can eliminate the human trait of biasness. AI will play its role in HR analytics removing the human biases from the process. AI can bring high level of automation to HR tasks. In the current scenario a lot of HR tasks are being automated and, in this line, it is expected that in future more tasks will be automated which will free employees for more strategic issues. The work structure will completely change with an increased involvement of AI. This will make human resources more creative and efficient. The re-skilling of people will provide cognitive diversity for the organization (Mathur, n.d.).

5 Conclusion

The above discussed scenario confirms the potential effect of AI on HRM practices. With the presence of industry 4.0, artificial intelligence will expand its applications and domains. The HR management should not ignore the transformation coming in its systems and practices due to the influence of artificial intelligence. The HR departments should be open to the potential change and re-design and re-equip themselves with relevant technologies and re-align themselves for embracing new challenges in future. They should not feel the fear rather find creative ways to take maximum benefit from AI techniques. Currently AI is considered as successful intervention for the basic HRM practices but its role in easing the complex tasks is yet to be explored. Moreover, the debate is still there that whether AI will contribute and affect in HRM or takeover many of its functions. So, there should be maximum focus on re-inventing the HR systems so that AI should complement the HR function and not substitute.

References

Aronson, J. E., Liang, T. P., & Turban, E. (2005). *Decision support systems and intelligent systems* (Vol. 4). Pearson Prentice-Hall.

Beattie, G., & Johnson, P. (2012). Possible unconscious bias in recruitment and promotion and the need to promote equality. *Perspectives: Policy and Practice in Higher Education, 16*(1) 7–13

Berk, A., & Kaše, R. (2010). Establishing the value of flexibility created by training: Applying real options methodology to a single HR practice. *Organization Science, 21*(3), 765–780.

Bondarouk, T., & Brewster, C. (2016). Conceptualizing the future of HRM and technology research. *The International Journal of Human Resource Management, 27*(21), 2652–2671.

Chen, C. H. (Ed.). (2013). *Pattern recognition and artificial intelligence.* Elsevier.

Chien, C. F., & Chen, L. F. (2008). Data mining to improve personnel selection and enhance human capital: A case study in high-technology industry. *Expert Systems with Applications, 34*(1), 280–290.

Davis, E., & Marcus, G. (2015). Commonsense reasoning and commonsense knowledge in artificial intelligence. *Communications of the ACM, 58*(9), 92–103.

Dickson, D., & Nusair, K. (2010). An HR perspective: The global hunt for talent in the digital age. *Worldwide Hospitality and Tourism Themes, 2*(1), 86–93.

Faliagka, E., Ramantas, K., Tsakalidis, A., & Tzimas, G. (2012). Application of machine learning algorithms to an online recruitment system. In *Proceeding International Conference on Internet and Web Applications and Services.*

Furneaux, B. (2012). Task-technology fit theory: A survey and synopsis of the literature. *In Information Systems Theory* (pp. 87-106). New York, NY: Springer.

Giotopoulos, K. C., Alexakos, C. E., Beligiannis, G. N., & Likothanassis, S. D. (2005, August). Integrating agents and computational intelligence techniques in e-learning environments. In *IEC (Prague)* (pp. 231–238).

Goodhue, D. L., & Thompson, R. L. (1995). Task-technology fit and individual performance. *MIS quarterly,* 213–236.

Griepentrog, B. K., Harold, C. M., Holtz, B. C., Klimoski, R. J., & Marsh, S. M. (2012). Integrial social identity and the theory of planned behavior: Predicting withdrawal from an organization recruitment process. *Personnel Psychology, 65,* 723–753.

Guchait, P., Ruetzler. T., Taylor, J. & Toldi, N. (2013). *Video interviewing: A potential selection tool for hospitality managers—A study to understand applicant perspective.*

Habeeb, A. (2017). *Introduction to artificial intelligence.* https://www.researchg ate.net/publication/325581483_Introduction_to_Artificial_Intelligence.

Hayes-Roth, F. (1997). Artificial intelligence: What works and what doesn't? *AI Magazine, 18*(2), 99–99.

HireVue. (2018). *Hirevue.* https://www.hirevue.com/products/video-interv iewing. Accessed 14 Dec 2019.

Jantan, H., Hamdan, A. R., & Othman, Z. A. (2010). Intelligent techniques for decision support system in human resource management. *Decision Support Systems,* 261–276

Johansson, J., & Herranen, S. (2019). *The application of artificial intelligence (AI) in human resource management: Current state of AI and its impact on the traditional recruitment process.* https://hj.diva-portal.org/smash/get/diva2: 1322478/FULLTEXT01.pdf.

Kaczmarek, T., Zyskowski, D., Walczak, A., & Abramowicz, W. (2010). Information extraction from web pages for the needs of expert finding. *Studies in Logic, Grammar and Rethoric, Logic Philosophy and Computer Science, 22*(35), 141–157.

Kanal, L., & Kumar, V. (Eds.). (2012). *Search in artificial intelligence.* Springer Science & Business Media.

Kaplan, J. (2016). *Artificial Intelligence: What everyone needs to know.* Oxford University Press.

Laha, A. K. (2019). *5 data-driven HR hiring and recruitment trends for 2019.* Retrieved December 16, 2020, from https://www.peoplematters.in/amp-rec ruitment-5-data-driven-hr-hiring-recruitment-trends-for-2019-21301.

Legg, S., & Hutter, M. (2007). A collection of definitions of intelligence. *Frontiers in Artificial Intelligence and applications, 157,* 17.

Martincevic, I., & Kozina, G. (2018). *The impact of new technology adaptation in business.* Varazdin: Varazdin Development and Entrepreneurship Agency (VADEA).

Mathur, S. (n.d.). *Artificial intelligence: Redesigning human resource management, functions and practices.*

McCarthy, J. (2000). Concepts of logical AI. In Minker J. (Eds.), *Logic-based artificial intelligence.* The Springer International Series in Engineering and Computer Science (Vol 597). Springer.

McCarthy, J. (2004). *What is artificial intelligence?* http://www-formal.sta nford.edu/jmc/whatisai/whatisai.html. Accessed 13 Dec 2019.

Merlin, R., & Jayam, R. (2018). Artificial intelligence in human resource management. *Artificial Intelligence in Human Resource Management, 119*(7), 1891–1895. ISSN: 1314–3395 (on-line version) http://www.acadpubl.eu/ hub/ Special Issue.

Minker, J. (Ed.). (2012). *Logic-based artificial intelligence* (Vol. 597). Springer Science & Business Media.

Montuschi, P., Gatteschi, V., Lamberti, F., Sanna, A., & Demartini, C. (2014). Job recruitment and job seeking processes: How technology can help IT. *It Professional, 16,* 41–49.

Newell, S. (2005). Recruitment and selection. In *Managing human resources: Personnel management in transition* (4th ed.). Blackwell Publishing.

Noe, R. A., & Kodwani, A. D. (2018). *Employee training and development* (7th ed.). McGraw-Hill Education.

Oxford Dictionary. (2019a). *Oxford dictionary: Definitions.* https://www.lexico.com/definition/artificial. Accessed 13 Dec 2019.

Oxford Dictionary. (2019b). *Oxford dictionary: Definitions.* https://www.oxford learnersdictionaries.com/definition/american_english/intelligence. Accessed 13 Dec 2019.

Sanders, M. R., & Murphy-Brennan, M. (2010). Creating conditions for success beyond the professional training environment. *Clinical Psychology: Science and Practice, 17*(1), 31–35.

Sierra-Cedar. (2018–2019). *HR Systems Survey White Paper, 21th Annual Edition.*

Stoilkovska, A., Ilieva, J., & Gjakovski, S. (2015). Equal employment opportunities in the recruitment and selection process of human resources. *UTMS Journal of Economics, 6*(2), 281–292.

Stuart, R., & Norvig, P. (2016). *Artificial intelligence: A modern approach* (3rd ed.). Prentice Hall Press.

Taylor, M. S., & Collins, C. J. (2000). Organizational recruitment: Enhancing the intersection of research and practice. In C. L. Cooper, & E. A. Locke (Eds.), *Industrial and organizational psychology: Linking theory and practice* (pp. 304–334). Oxford, U.K.: Blackwell.

Tecuci, G. (2012). Artificial intelligence. *Wires Comp Stat, 4*, 168–180. https://doi.org/10.1002/wics.200.

Upadhyay, A. K., & Khandelwal, K. (2018). Applying artificial intelligence: Implications for recruitment. *Strategic HR Review, 17*(5), 255–258.

Ved, S., Kaundanya, N. S., & Panda, O. P. (2016). Applications and current achievements in the field of artificial intelligence. *Imperial Journal of Interdisciplinary Research, 2*(11), 932–936.

Verma, R., & Bandi, S. (2019). *Artificial intelligence & human resource management in Indian IT sector.* Available at SSRN 3319897.

Wall, T. D., & Wood, S. J. (2005). The romance of human resource management and business performance, and the case for big science. *Human Relations, 58*(4), 429–462.

Zuboff, S. (1985). Automate/informate: The two faces of intelligent technology. *Organizational Dynamics, 14*(2), 5–18.

Industrial Revolution 4.0: Transformation of Job Market

Girija Narasimhan, Reshmy Krishnan, and Ashok Krishnan

1 Introduction

"The Future of Jobs: Employment, Skills and Workforce Strategy for the Fourth Industrial Revolution, a host of developments have moved the world of work into the '4th Industrial Revolution' which will fundamentally alter the way we live, work, and relate to one another" [January 2016 report from the World Economic Forum (WEF)]. IR 4.0 is the subsystem of the fourth industrial revolution which started in Germany. It refers to the innovation changes toward automation and data exchange in manufacturing technologies, processes such as cyber-physical systems (CPS), Internet of Things (IoT), cloud computing, cognitive computing, and artificial Intelligence. The IR 4.0 is implemented in various domains such

G. Narasimhan
University of Technology and Applied Science, Muscat, Sultanate of Oman

R. Krishnan (✉)
Muscat College, Muscat, Sultanate of Oman
e-mail: reshmy@muscatcollege.edu.om

A. Krishnan
Central Queensland University, Rockhampton, QLD, Australia

© The Author(s), under exclusive license to Springer Nature Singapore Pte Ltd. 2021
N. R. Al Mawali et al. (eds.), *Fourth Industrial Revolution and Business Dynamics*, https://doi.org/10.1007/978-981-16-3250-1_13

as transportation, health care, robotics, manufacturing, autonomous shipping yard, etc. with technological revolutions such as the Internet of Technology (IoT), cloud storage, artificial intelligence, etc. The business process is changed in a smarter way with the innovations of IR 4.0. Human-to-Machine communication evolving toward Machine-to-Machine (M2M) communication and stepping forward to the smart world. Smart industry revolution required machine learning and data analytical ability, which is part of Artificial Intelligence. In the future, smart technology will be vital everywhere such as healthcare, education, customer service, manufacturing, construction, transportation, military, etc. Smart technology plays a sufficient percentage of ratio in employment, which leads to intellectual challenges for job seekers. The significant unreciprocated tasks of IR 4.0 is

- How smart technology, CPS, or AI can benefit society?
- What type of risk in the near future on educational skills?
- Whether smart technology rides the society or society controls the technology?

1.1   Through the Revolutions

During the second half of the eighteenth century and the beginning of the nineteenth century, a period of development happened mainly in rural societies of Europe and America. That development called the industrial revolution, which transforms rural society into urban. With the introduction of new techniques and machines, mass production of textiles, iron, and other products happened. During the '30s and '40s of the eighteenth century, using water and steam power to mechanize production, the industrial revolution began in the rest of the world. The advancement of communication methods also happened during the second half of the eighteenth century. In 1602 in Amsterdam, the first stock exchange was established. Next one in New York in 1790.

At the end of the nineteenth century, new technologies like electricity, gas, and oil initiated the development of the combustion engine for its full potential. The steel and chemical industry began growing with the development of synthetic fabric, dyes, and fertilizer. The invention of the telegram and the telephone lead to the growth of communication during this period and transportation methods advanced with plane and automobile inventions.

The third industrial revolution started in the second half of the twentieth century with the development of nuclear energy, which causes the rise of electronics. Transistors and microprocessors, telecommunications, and computers emerged because of this. Space research and biotechnology are the other two new technologies developed during this time. The era of high-level automation in production happened with the inventions of programmable logic controllers (PLCs) and robots.

1.2   Technological Revolution

Technologies such as artificial intelligence, the Internet of things, Blockchain, augmented reality, robotics, Nanotechnology, energy storage, cloud computing, quantum computing, and 3-D printing are transforming the way humans create, discuss, and spread values. Transformations of organizations, industries, and individuals are occurring here. The future world is indebted to these powerful new technologies. Social transformation such as the way of communication, learning, analyses, decision making relate to one another and entertain ourselves is the result of all these emerging technologies. In this revolution, human beings are getting the opportunity and responsibility to structure it and apply it rather than being the victims of the technologies.

Adoption of cyber-physical systems like the Internet of things (IoT) and Internet of Systems (IoS) are the major happening in IR 4.0. The IoT is the network of smart devices in which each device are interacting with other devices. IoS happens when Business systems gather data from IoT to make independent decisions in various business domains such as marketing, strategies, sales, production, etc.

Artificial Intelligence is making impressive progress in recent years with the combination of computing power and data. Its involvement is all around us from self-driving cars, intelligent robots, drones, mobile supercomputing, virtual assistants, business support systems to emotional support systems. As a result of these emerging technologies, there are changes occurring at an exponential speed in daily life.

2   Literature Review and Analysis of Related Work

2.1   Origin of IR 4.0

Over the twenty-first century, a fourth industrial revolution happened in social, political, cultural, and economic domains recognized. After the digital technology enhancement in the third revolution, IR 4.0 represents the combination of digital, biological, and physical innovations. IR 4.0 influence change inhumane life and work. The extraordinary technologies merge all manual, global, and biological works and lead to social development. IR 4.0 gives hope and confidence toward the countries development, organizational growth, and stability in society. It is the opening to serve all domains of human beings like leaders, policymakers, middle-level employees, skilled employees, and promoting a human-centered future.

The fourth revolution occurs at an exponential pace rather than linear. The breadth and depth of this herald the transformation of thorough systems of production, management, and governance.

2.2   The Impact of Fourth Industry Revolution 4.0

We can affordably gain the digital world now a day due to the emergence of the fourth revolution. The personal life of an ordinary person has become more progressive and delightful, because of modern technology products and their services. For example, booking a flight, making payment, purchasing a product, etc., can be done remotely. Supply-side miracle has happened because of technological innovations. Communication methods and transportation have become less expensive. The global supply chains and logistics become more effective, the trading cost reduced, and a new horizon opened in marketing. Ultimately, economic growth emerges.

The revolution may cause wider inequality because labor markets are disrupted because of automation. The replacement of employees by machines leads to the gap between returns to capital and returns to labor. Talent has become a critical factor of production than capital therefore segments like low-skill/low-pay and high-skill/high-pay were generated.

3   Research Methodology

3.1   Role of New Technology in Employment Skills

As per the World Economic Forum report 2018, emerging technologies like Big Data analytics, IoT, Machine learning, Cloud computing, and Digital trade, etc., will play business processes in the future. The emerging technologies are absolutely in the business environment as per the Data Quest blog "10 high-Paying Jobs That Require a Knowledge of Data Analytics." A comparison of the job vs. the average salary of topmost employment skill as per report top most salaries are given to IT System analyst $68.80 and Data Analyst salary ranging from $61 to $73. It reflects the job market is extremely welcoming to industrial revolution 4.0. Nevertheless, indirect job opportunities like training institutes, certificate courses, book authors, and content writers are needed for bridging corporate employment. Qualified professionals trainers are essential, which also produce more employments in academia.

3.2   Occupational Transformation

New technology not only decides the new expertise set. It also points out some skills which vanished or transformed based on industry requirement. As per Mckinsey Manichean guessing game discussion, the transformation of automation of technologies with employment skills was categorized into three groups such as highly susceptible, less susceptible, and least susceptible.

There are two sectors where low automation exists health care and education. Both of these sectors demand knowledge-based expertise like handling people, creative thinking, and decision-making. To date, automation is replacing the majority of the physical working skill set. At the same moment, Artificial Narrow Intelligence (ANI) systems like chess-playing Deep Blue and Deep Mind AlphaGo are smart technologies that are more progressive than the human brain. These technologies are remarkably brilliant than the human brain. IR 4.0 not exclusively aims at industrial automation but also pointing toward machine learning and deep learning concepts. The hypothetical point of this technology is described as intellectual exposition. And is also known as the technological singularity. As per McKinsey's report, various sectors are increasingly using AI-based Technologies. These sectors are categorized into three least susceptible using 7–14% and less susceptible using 20–25% and

Table 1   Various sector using AI based technology

Sectors	Technology	Description
Health care	Gauss Surgical computer visions	Nursing while obstetric hemorrhage (MOH) bleeding during Caesarean cases
Transportation	Waymo computer vision technology	Self-driving car
Oil and Gas	Osprey's intelligent visual monitoring	find leaks, safely monitor and strengthen security especially remote field
Agriculture	Aerial Phenotyping ("AP") technologies	Crop development monitoring system
Education	*Automated essay scoring* (AES)	Using Natural language processing grading the essay
News and Media	Lynx Insight- AI based Journalist	Used by Reuters, especially to focus the stock market data
Designing	Generative Adversarial Network (GAN)	Generating photographs and cartoon design

Source Compiled by the Authors

highly susceptible 64–78%. The least susceptible job like managing others and applying expertise and less susceptible jobs are stakeholder interactions and a highly susceptible job like data processing and predictable physical work.

Hybrid thinking and Artificial Super Intelligence (ASI) technologies are moving toward to read human brain concepts. Table 1 listed out all the technology that helps work culture in a timely fashion. Table 1 information indicates that how automation replacement will occur in the future. It also specifies that automation is not only in the manufacturing sector. In the forthcoming, automation plays a sufficient percentage of all the sectors in the industry.

3.3   Digital Transformation

As per Deloitte's insight, organizations transformed toward real-time data and intelligence operations through Industry 4.0. There occurs a transfer toward data and software in organizations. The initial cost is needed for purchasing an intelligent machine. The advantages of these revolutions are more consumer benefits, reduction in power consumption, and improvement in comfort. The business developed, as the consumers are happy to pay for data-driven insights. Challenges need to address to make equilibrium between the legacy businesses and modern business models with innovative offerings. As per Deloitte global surveys, digital transformation remains the utmost strategic priority in the upper management of significant organizations. It is evident in the survey that inconsistencies are observed in strategy, supply chain transformations, talent readiness, and drivers for investment.

Strategy paradox: Employees of organizations should be aware of business growth and explore the strategic possibilities with digital transformations rather than acknowledge strategic importance.

Supply chain paradox: The notable point that the utmost priority in digital transformation is the supply chain. The lower or middle management is not fulfilling a proper role in decisions.

Talent paradox: It pointed out that identifying, training, and retaining the right talent as top management is essential in digital transformation.

Innovation Paradox: The primary cause of digital transformation initiatives observed productivity improvement and objectives of the operations. Deloitte survey suggested a broader term of innovative technologies for near-term business operations. So that increased desire for innovations and strategic focus will be the positive return on investment.

German-based multinational automation company Festo demonstrates how the transformation of IR 3.0 to IR 4.0. Festo Didactic is a global trainer in technical fields such as pneumatics, hydraulics, electrical engineering, production technology, mechanical engineering, mechatronics, CNC, HVAC, and telecommunications. Festo Didactic alliance with SAP University developed curriculum, which fulfil theoretical knowledge combination of practical experience in smart IR 4.0 initiatives.

Qualitative change in the labor market have occurred with skill-based technological change. Hence skill demand and educated workers need to address (Hartmann & Bovenschulte, 2013). Job polarization has happened in the labor market because of the demand for a highly skilled job and high wage occupations. Researchers found that computer-intensive industries demand more skill upgrading (Singh et al., 2018).

Noteworthy analysis has been conducted worldwide on the skill gap because of the use of modern technologies. The transition from employment to unemployment increased in France, West Germany, and the USA. To address this issue another approach-routinization- initiated in which tasks focused on wages (Renjan & Brown, 2018). As per this hypothesis, routinization may occur both in low and high skilled jobs. As a result, new technologies compliment or support low or high-skilled jobs. In this case, it is advisable to mention details of routine and non-routine tasks. These tasks are classified into two categories like abstract/cognitive and manual tasks. The cognitive or analytical tasks demand soft skills like problem-solving, creativity, and analytical skills for occupations of all domains.

The Manpower group solution is a Fortune 500 company in the USA that conducted the "Talent shortage Survey." The survey report states that massive companies are facing a talent shortage of 67% globally. The survey has 12 years of research data. This survey was conducted on 39,195 employees in 43 countries.

4   RESULTS

It observed a lack of applicants with experience is the challenging factor, which causes a skill gap. The key factor to point out, is that the applicant must have the skillset and emerging technologies required for the future job market.

A strategic plan to be developed to recognize the digital skill capacity of the resources and how automation will transform the organization. The work organization can review to identify the changes required as automation affects the entire business. The management layer should ensure to adapt to the challenging environments and handle risks due to this. The business sector can associate with national and international higher education institutions. They involve in suitable student project activities to retrieve distinct ideas for the organization. The current resources can upskill to adopt recent changes. Investing in research and development like a collaboration with a research organization helps to gather innovative practices. The business organization should be aware of the ministry initiatives and supports for digitization and utilize them.

5   CONCLUSION

IR 4.0 causes digital and occupational transformations in the world. The revolution may cause vast inequality because of the labor market disruption due to automation. The studies show the job market is excessively welcoming to industrial revolution 4.0. At the same instant, indirect job opportunities require bridging corporate employment. Qualified professional trainers are essential, which also produce more job opportunities in academia. As per the discussion, the transformation was categorized into three groups like highly susceptible, less susceptible, and least susceptible. The studies shows that the industrial revolution 4.0 is excessively welcomed in job market. Consequently, more consumer benefits, reduction in power consumption, and improvement in comfort represent the advantages of this revolution. The survey revealed that inconsistencies observed in strategy, supply chain transformations, talent readiness, and drivers for investment. Job polarization has happened in the labor market because of the demand for extraordinarily skilled jobs and high wage occupations. The business strategy reveals that the top management layer should ensure to adapt to the challenging environments and handle risks due to this. Innovative practices can be brought through investing in research, collaborations with research organizations, awareness of current ministry support and initiatives.

References

Amiron, E., Latib, A., & Subari, K. (2019). Industry revolution 4.0 skills and enablers in technical and vocational education and training curriculum. *International Journal of Recent Technology and Engineering (IJRTE), 8*(1C2), 484–490. ISSN:2277–3878.

Autor, D. H., Levy, F., & Murnane, R. J. (2003). The skill content of recent technological change: An empirical exploration. *The Quarterly Journal of Economics, 118*(4), 1279–1333.

Goldin, C., & Katz, L. F. (2007). *The race between education and technology: The evolution of US educational wage differentials, 1890 to 2005* (No. w12984). National Bureau of Economic Research.

Hartmann, E., & Bovenschulte M. (2013). Skills needs analysis for "Industry 4.0" based on road maps for smart systems. In *SKOLKOVO Moscow School of Management and International Labour Organization*. Corpus ID: 199480874.

Lee, J. W., & Wie, D. (2015). Technological change, skill demand, and wage inequality: Evidence from Indonesia. *World Development, 67*, 238–250.

Renjan, P., & Brown S. (2018). *Preparing tomorrow's workforce for the fourth industrial revolution for business: A framework for action*. Global Business Coalition for Education, Deloitte.

Singh, D., & Tilak, G. (2018). Industry 4.0—4th rising industrial revaluation in manufacturing industries and its impact on employability and existing education system. *Pramana Research Journal, 8*(11), 161–169. ISSN NO: 2249–2976.

Sniderman, B., Mahto, M., & Cotteleer, M. J. (2016). *Industry 4.0 and manufacturing ecosystems: Exploring the world of connected enterprises.* Deloitte Consulting.

Spitz-Oener, A. (2006). Technical change, job tasks, and rising educational demands: Looking Outside the wage structure. *Journal of Labor Economics, 24*(2), 235–270.

Wu, C. H., & Luh, D. B. (2018). Myths surrounding innovation design in the industry 4.0 era. *Management Review, 37,* 123–136.

Industry 4.0: The Human Resource Perspective

R. Nirmala and Neha Chitte

1 INTRODUCTION

The interface of technology and humans can be witnessed in different walks of life. Many emerging technologies like Artificial intelligence, Block chain, Internet of Things, etc. have made its way in human lives. The main objective of any new technology is basically to ease day-to-day life. One such example of technology is robotics. Recently, a restaurant in Chennai has successfully introduced robots as waiters. It is surprising to see the job that was considered to be completely person-centric being smoothly performed by such new "techno waiters" and also appreciated by the customers. Such reduction of human element in businesses processes has opened a new question, can machines replace human beings completely and if yes how? Is it feasible to install systems to perform range of tasks to reduce cost, time, and improve quality by completely bypassing humans? Probably not. However, today a smart tag is very

R. Nirmala
Goa Business School, Goa University, Panaji, India
e-mail: nirmala@unigoa.ac.in

N. Chitte (✉)
Goa Business School, JRF Scholar, Goa University, Panaji, India
e-mail: mba.neha@unigoa.ac.in

N. R. Al Mawali et al. (eds.), *Fourth Industrial Revolution and Business Dynamics*, https://doi.org/10.1007/978-981-16-3250-1_14

269

common on different products and systems, including phones, televisions, vehicles, and even whole cities. SMART means "Self-Monitoring Analysis and Reporting Technology" Thus smart devices, smart systems, and smart infrastructure collectively can create a new eco-system of technology which significantly differs from earlier ones.

Once upon a time, not so long ago, "Business transformation" was considered a fashionable term and "Organisational Development/Business Process Re-Engineering" considered periodic interventions to be introduced by progressive organizations. But today, the applications of technology in businesses are far reaching and much sought after to ease business processes.

It is said that the World has witnessed three Industrial revolutions till recently.

1.1 The first industrial revolution started in sixteenth century was dominated by hydraulic and steam power. This revolution has its origin in Europe and covers the period from 1760 to somewhere in 1840. It was a complete transformation of manufacturing processes from hand production methods to machine production. Steam power, waterpower, mechanical power, mechanization, new chemical manufacturing, iron production, were the main features of this revolution. The main industry focussed and benefited from this revolution was textile industry. Hence it brought about significant socio, economic, and political changes. There was an emergence of working population, a significant increase in labor force. The job opportunities were generated to a great extent, but the working conditions were strict with long hours of labor set in accordance with the pace of machines. Comparatively, the wages earned by the labor were 20–40% less to lead a decent life. Other issues like unhygienic working conditions for women, children, and through organization of workers, trade unions, and even strikes at later period of time. The phenomenon of machine and human interaction then became the basis for next revolution. Child labor was prevalent then. On the other hand, labor found representation.

1.2 The second industrial revolution used electric energy and the assembly production lines. It was basically dominated by mass production. It began in late nineteenth century to early twentieth century; the phase of rapid industrialization. The famous quote by Henry Ford about the Ford T model car, "You can have any color as long as it is black" indicates the same. Mass production without product differentiation is the feature of this revolution. The main sectors dominated by this revolution were iron, steel, coal, petroleum, paper making, railways, automobiles, etc. the productivity improved significantly. The prices of goods decreased significantly. Hence it caused lot of upheavals in industry giving rise to unemployment. Labor was displaced by machines and many factories, ships became obsolete. On the other hand, there was division of labor which made both skilled and unskilled labor more productive. This led to the creation of professional class and also considerable decrease in child labor.

1.3 The third industrial revolution included computers and information technology. It was dominated by electronics and nuclear technology. The two main components of electronics i.e., transistor and microprocessor gave emergence to telecommunications. This revolution focussed on microelectronics and flexible production. A variety of products were manufactured with programmable machines. But the limitation was it did not have flexibility concerning production quantity. There was shift in the kind of jobs. The knowledge of handling computer became a paramount skill which even created havoc among the aged employees. Hence need for retraining, voluntary retirements, etc. arose during this phase.

1.4 The fourth and current Industrial revolution involves ICT (Information and Communication Technology) wherein there is interface between cyber physical systems with automation and decentralized control. This leads us to "smart" manufacturing (Fig. 1).

2   What is Industry 4.0?

Industry 4.0 is basically the subset of fourth industrial revolution focussed on industry. Whereas fourth industrial revolution includes areas classified other than industry and includes range of fields that impact society like "Smart agriculture." The three fundamental technological drivers of this revolution are digital, physical, and biological technologies. Industry 4.0 and fourth industrial revolution are used inter-changeably. Industry 4.0

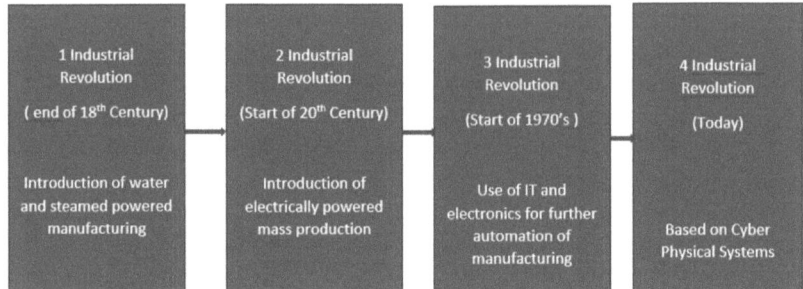

Fig. 1 Stages of Industrial Revolutions (*Source* Created by Authors based on Progression of Industrial Revolutions characterized by different technologies as proposed by Kagermann et al. [2013])

has its origins in Germany, one of the most competitive and advanced industries in the world. In 2011, at the Hanover trade fair, the term "Industry 4.0" was made known in public domain, as the name for the joint initiative of the representatives of business, policy, and science to strengthen the competitiveness of German industry. The Federal Government of Germany then undertook it as an important initiative "High-Tech Strategy 2020 for Germany."

Industry 4.0 includes the integration of cyber physical production systems (CPPS), wherein there is interface between humans and computer. In Industry 4.0 production floor, systems are integrated with ICT components. These are autonomous systems involved in decision making at real time which make use of machine learning algorithms and results based on past behavior.

Industry 4.0 includes vertical and horizontal integration of production systems connected on real time basis. In a smart factory, the tools like cloud computing, big data analytics are used as a source of data to kick start production. The production is based on this data which includes data from various stakeholders like customers, suppliers, competitors, etc. Then the digital to physical conversion takes place through human–machine, machine–machine, and human–human collaboration. Tools like artificial intelligence, machine learning, robotics, are used in this step. Similarly, both operations and control processes are performed using advanced tools like virtual reality, addictive manufacturing, etc. The interoperability of products in a smart factory takes place through sensors.

The processes and systems are decentralized, modular, autonomous, self-optimizing, and self-organizing in nature. Thus, a smart manufacturing configuration involves both smart machines and smart humans.

3   Definitions of Industry 4.0

3.1 As per Schumacher, Erol and Sihn (2016) Industry 4.0 is surrounded by a huge network of advanced technologies across the value chain. Service, Automation, Artificial Intelligence Robotics, Internet of Things, and Additive Manufacturing are bringing in a brand-new era of manufacturing processes. The boundaries between the real world and virtual reality are getting blurrier and causing a phenomenon known as Cyber-Physical Production Systems (CPPS).

3.2 Lu (2017) claims, that "Industry 4.0 can be summarized as an integrated, adapted, optimized, service-oriented, and interoperable manufacturing process which is correlate with algorithms, big data, and high technologies."

3.3 From management perspective, Industry 4.0 is defined as the integration of Internet of Things technologies into industrial value creation enabling manufacturers to harness entirely digitized, connected, smart, and decentralized value chains able to deliver greater flexibility and robustness to firm competitiveness and enable them to build flexible and adaptable business structures, acquiring the permanent ability for internal evolutionary developments in order to cope with a changing business environment as the result of a purposely formulated strategy implemented over time (Piccarozzi et al., 2018).

4   Components of Industry 4.0

4.1   Horizontal Integration

From the operational perspective, horizontal integration occurs when company's focus is primarily on its core competencies and its end-to-end value chain is established through partnerships. While with regard to business, horizontal integration means integration of organizations that targets the similar customer base with different products or services. In

Industry 4.0 context, it means there is a connected network of cyber physical systems which involves high level of automation, that is flexible and operationally efficient. It brings the concept of a new type of worldwide value chain networks.

4.2   Vertical Integration

From operational perspective, in vertical integration value chain is maintained within the in house as much as possible—from product development to sales. While in the context of business, it means integration with the companies that bring avenues of reduced manufacturing cost, better market opportunities, etc. In Industry 4.0 context, it refers to integration of all processes within the organization i.e., from production and marketing. Strategic and tactical decision making is enhanced due to the free flow of data.

4.3   End to End Engineering

This includes integration of entire value chain, to assist in customized production. In this the demands of customers, product design and development, recycling, maintenance is taken into consideration. Thus, there is free flow of information and customized product.

5   Characteristics of Industry 4.0

Automation drives Industry 4.0, which includes digitized factory and digitized products. Nevertheless, Industry 4.0 is still an emerging area hence the academic field finds it difficult to distinguish between its key features.

There are 10 characteristics of Industry 4.0:

- Artificial Intelligence—It is the way of making a computer, a computer-controlled robot, or software think intelligently, in the similar manner the intelligent humans think.
- Cloud Computing—These are basically data centers available to many users over the Internet. Cloud computing makes available computer system resources like data storage and computing power.

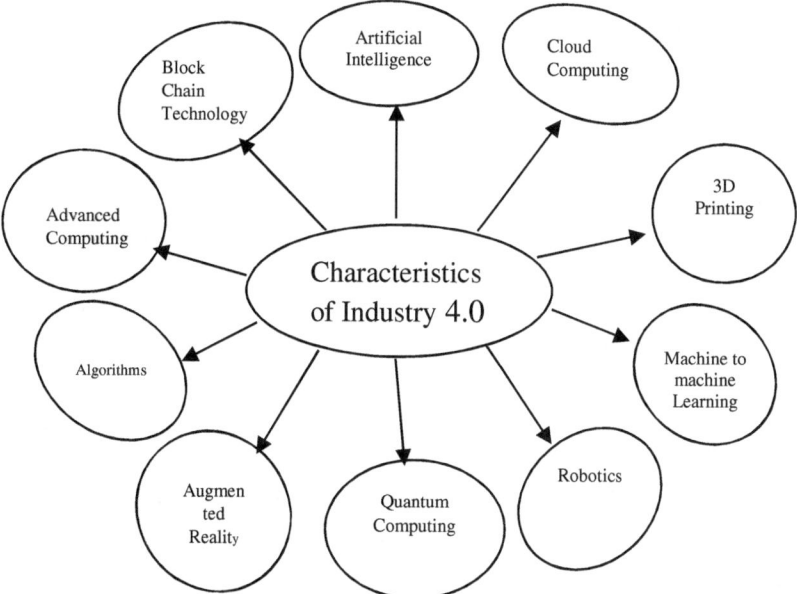

Fig. 2 Characteristics of Industry 4.0 (*Source* Created by authors based on Morteza Ghobakhloo, 2018)

- Block Chain Technology—Block chain is a decentralized technology. A global network of computers uses block-chain technology to jointly manage the database. It is also used as a distributed ledger.
- Machine to Machine Learning—Machine learning is an application of artificial intelligence (AI) that provides systems the ability to automatically learn and improve from experience without being explicitly programmed.
- 3 D printing—3D printing is any of various processes in which material is joined or solidified under computer control to create a three-dimensional object, with material being added together, typically layer by layer.
- Robotics—Robotics deals with the design, construction, operation, and use of robots, as well as computer systems for their control, sensory feedback, and information processing.
- Quantum Computing—It is computing using quantum mechanical phenomena, such as superposition and entanglement. Quantum

computation uses quantum bits, qubits unlike binary bits used in computers.

- Augmented Reality—A technology that superimposes a computer-generated image on a user's view of the real world, thus providing a composite view.
- Algorithms—Algorithms can perform calculation, data processing, and automated reasoning tasks. A process or set of rules to be followed in calculations or other problem-solving operations, especially by a computer.
- Advanced Analytics—Determined to create business perceptions from an accumulation of data by pinpointing patterns and interdependencies (Fig. 2).

6   Relation of Industry 4.0 to Human Resource Management

Industry 4.0 is perceived to be a disruptive innovation but a necessary evil, indeed poses many impacts on the way humans work in an organization. Though it provides for many opportunities like performing complex task swiftly through automation, its real interaction with humans at work needs a thorough analysis.

The relationship between humans and machines is changed in Industry 4.0 environment. The repetitive, monotonous, dull activities are performed by machines. While activities like strategic decision making, agile working, smart working is expected by the humans of this era.

But another perspective in this area is the growing prophesies that machines would replace humans not only in routine jobs but even in the cognitive ones. Thus, it creates questions like what would be the new number and types of jobs created and what would be the jobs in which humans can excel?

This indicates that HRM 4.0 needs to shift its focus on tasks which are non-routine in nature, based on evidence, value addition, and derive its inspiration from science. Work, creativity, and technology need to go hand in hand. Thus, Smart Human Resource (SHR 4.0) becomes a promising solution. It is a recent concept evolved as a part of Industry 4.0 HR domain, wherein the focus is on innovations, adoption of recent technologies to solve HR problems. It is basically the digital transformation of HR functions like recruitment, training, performance, etc. based on "People's sciences."

HR department in any organization is the one responsible for everything in and out about its workforce. Beginning from the recruitment till the exit of an employee, the role of HR department is very important. But today HR departments across the organizations are seen playing majorly the operational role with the obsolete technology and traditional skill sets. On the other hand, the exponential growth of technology, transmission of Big data, and Artificial intelligence (AI) demands data analysis precisely and on real time basis. Also, along with the technological transformation, the generation of employees entering the workforce is also witnessing a significant drift. Gen Y (birth year between 1980 and 2000) employees would be half of the workforce by 2020. Gen Y and Gen Z (birth year after 2000) are the ones grown amidst the technology (internet, Mobiles, social media, smart gadgets). Thus, these current or potential employees have different set of expectations from their employers like flexible working, work-life balance, anytime anywhere communication, real time feedback, free and open work culture, opportunities for career growth through data driven technologies, etc.

Hence SHR 4.0 with its data driven technologies and its new technologically oriented employees indeed has the capacity to transform the entire HR systems and processes.

The various functions of HRM can be analyzed through the prism of Industry 4.0 as follows:

6.1 Recruitment and Selection

As smart phones are equipped with the precise app settings Gen Y and Gen Z employees can be tactically reached using targeted advertisements based on their profiles and preferences saved in those settings. Using tools like Artificial intelligence, Big data, Cloud Computing, and Block chain technology the resumes can be sorted out with those fulfilling the job description and job requirements. Interviews can be conducted using faster data networks like 4G and 5G and AI chat bots can help in real time assessment of the candidate's responses. Finally, the new joiners can be guided through the Virtual Reality (VR) and Augmented Reality (AR) . Similarly, soft computing can assist in personnel assessment and selection as well as job assignment of personnel. Machine learning is another tool that has shown significant rise in internal hiring from 4 to 60% in Canada and US. Hence such talent on boarding performed using advanced tools promises precision and effectiveness in hiring.

6.2   Training and Development

The next important step after on boarding is to enhance the knowledge, skills, and abilities of employees through appropriate training and development methods. This requires proper assessment of competencies like technical, methodical, social, and personal competencies. Artificial Intelligence can help in identifying the skill gaps and help in competency mapping. Similarly, Virtual trainings can be imparted to the Gen Y and Gen Z employees. Also, the technology of cloud computing would enhance the speed and accuracy of training needs assessment and career development. In this dynamic environment the need for retraining is likely to arise which can be done using such data driven technologies. Thus, Human capital should be adapted, and digital skills are compulsory.

6.3   Performance Management

In traditional HR practices, performance management is once in a year ritual wherein the one size fits all approach is followed. But for transformation of HR to SHR 4.0, performance targets need to be set up on individual basis. Thus, artificial intelligence can assist in setting up individual performance goals. Promotions can be based on the KPI (key performance indicators) rather than only on the basis of seniority. Also using cloud computing and big data analytics, performance feedbacks can be managed to derive appropriate performance indicators and results. Similarly, the wearable IOT (Internet of Things) technology equipped with its "sensitive digital electronic-network built-objects" can help monitor and measure data to improve performance. This may also assist in off boarding, by tracking the high and low performers' thereby giving internal opportunity for high performers leaving the organization.

6.4   Motivation

The advent of sophisticated tools and technologies such as robotics solving complex tasks may Undermine the role played by humans thereby affecting their morale and motivation. But studies point out that the adoption of technology can have both negative and positive impact on motivation. Motivation in context of Industry 4.0 can be studied as:

1. Process motivation factor-It is achieved by properly formulated demands process operation, its inputs and outputs. This system assumes the correct and straightforward e-communication between the workers concerned by production planning and organization processes. Here the role of mutual communication among the workers and their teamwork plays an important role.
2. Product motivation factor—In Man-Man systems the worker is responsible for particular tasks. Also, it's necessary to have required competence to spark necessary changes without the occurrence of conflicts in the integrated systems. Thus, motivated staff plays an important role and are source of innovations in complex e process environment.
3. Personality motivation factor—Here the rotation of employees in the manufacturing process, skilling, providing avenues for cooperation, knowledge sharing and learning, teamwork is considered important parameters.

Thus, the core motivation for implementation of Industry 4.0 concept is relevant data and information available in real time for flexible process management.

6.5   Employee Welfare

The compensation management can be done from the database available. Similarly, employee health and wellness can be monitored using the wellness applications and smart wearable devices. Such devices are based on the technology of Internet of Things (IoT) . This fitness parameters can be tracked on real time basis and could reduce the number and frequency of sick leaves.

6.6   Reward Management

In context of Industry 4.0 rewards should be the basis of motivation and a retention tool for human capital. Human Resource can be tranformend into human capital when it is acquainted with Internet of Things (IoT), human-machine interations, and well versed in understanding network systems.

7   Advantages of Adopting Industry 4.0 Tools in HRM

a) Customized cost and time effective HR operations.
b) An improved employee-employer experience.
c) Helps for mass production without increasing overall human resource costs.
d) Increasing flexibility, open, innovative, and better working environment.
e) Sustainable use of energy and resources for better talent management.
f) Leaner HR departments focusing on key strategic areas with precision.
g) Efficient talent management.
h) Shift of focus from operational to strategic policy formulations and implementations.
i) May reduce in employee attrition by promoting the principles of job fit theory.
j) Improvement in job satisfaction due to enhanced learning, career development, and effective work-life balance.

8   Challenges in Implementation of Industry 4.0 with Respect to HRM

Numerous studies have pointed out the changing dynamics of HRM in Industry 4.0 environment. The types of jobs, skills, competence, training, etc. considerably need a relook in the new workplace equipped with cyber physical systems.

8.1   Change Management

Change is the law of the nature. While resistant to change is omnipresent, but the intiators of new technology are humans themselves. Though top management is primarily associated with the transition, every worker needs to be taken into consideration for sustainable change management.

It thus challenges the traditional job design. These challenges require both people and organizational change. Also, there is a need to deal with increase work stress, effect on critical thinking ability due to the use of machines. Thus, it provides for another challenge for HRM domain to opt for digital means of managing, organizing, and leading change. Additionally, overcoming existing work culture, and managing multi-generational employee expectations is another factor.

8.2   Decentralization and Standardization

As Industry 4.0 involves free flow of information on real time basis there is decentralization wherein multiple stakeholders contribute through their insights. Thus, this environment demands real time, analytical, and complex role of humans. As manual and repetitive tasks are automized, humans are more required in decision making, R&D, and control. Hence this poses challenges like resistance from workers, threat to intrinsic motivation of workers, fear of losing current position, etc. Hence continuous and comprehensive training plays an important role in maintaining the morale of workers.

8.3   Education, Training

New working sphere demands new skills and knowledge from the worker. Conventional rote learning method does not suffice the need of technologically advanced job roles. Thus, the application of theoretical knowledge, and innovation is the need of the hour. Humans have to be ready to adapt to the constant changes taking place in the market, technology, systems, etc. Thus, company training and development measures ensure smooth shift from lean learning factory to Industry 4.0 learning factory.

8.4   Human Machine, Human–Human and Machine–Machine Interactions

Humans perceive machines as a threat to their jobs. But the manual repetitive jobs would be taken over by machines leaving the analytical, complex tasks to humans. Hence the role of humans does not become redundant rather it becomes more interactive. This interaction may seem unpleasant initially mainly for the older workers, but the ones who know how to deal with machines and optimize their work would survive. The Human–Human interaction in Industry 4.0 is done via special social networks which give different social aspect to the company. The machine–machine interaction is based on automation wherein the systems take decisions independently are expected to warn humans of the needed requirements through controlling. Hence the role of humans remains important to ensure that the processes do not do more harm than benefits.

8.5   Decision Making

In industry 4.0 work environments, decision making is a decentralized phenomenon unlike traditional method. Thus, quick and prompt decisions have to be made. The data derived from various systems is complex and have to be incorporated wisely to arrive at decisions. This is analytical human tasks which becomes more complex in Industry 4.0 scenario. The data interpretation also needs to be accurate. Thus, the competence of using the DSS, the ease of DSS systems to simplify the complex tasks given to humans becomes a challenge.

9   MANAGERIAL IMPLICATIONS

a) *Changes in the organizational structure*—Industry 4.0 demands some change in overall organizational design like a flat hierarchy-based agile organization structure from the traditional model.

b) *Decentralization of power*—A shift from top-down approach to a decentralized approach due to vertical and horizontal integration.leadership styles- A less of an authoritarian and more inclined towards participative and delegative leadership style is preferable in Industry 4.0 management settings. Leadership should promote more learning and innovation culture.

c) *Efficient grievance redressal mechanism*—Management must be vigilant towards the conflicts that may arise due to multi generation employees and the disruption caused by the advent of new technology.

d) *Strategic Management*—The management needs to emphasis upon the technology oriented, practically feasible, human resource friendly, sustainable policy formulations and implementations. Management needs to change its focus on strategic workforce planning.

e) *Implication on HR department*—Reduction in HR team size and increased opportunity for HR department to focus on·more strategic areas.

10   OVERALL IMPACT OF INDUSTRY 4.0

The implementation of Industry 4.0 has both positive and negative impact on human resource management. The changes in the job structure, skills, and competencies have both positive and negative dimensions.

10.1   Positive Impacts

Adoption of Industry 4.0 has positive impact on human resource productivity. It means with the use of advanced tools and new technologies works are done at a faster pace as compared to the traditional techniques and tools. For e.g., robots have better working ability and high potential of doing work as compared to humans. Robots can do jobs faster and better. Along with the time factor procurement and maintenance of machines is cost effective than hiring, training, and retaining the employees. Similarly in the countries with high number of ageing populations, machines are better alternative as compared to humans. The implementation of sophisticated machines gives competitive edge to the organization. Due to new technology the training needs arise thus maintaining a well-trained, professionally skilled, and technologically competent workforce in an organization. In case of MSME's, assistance from the government agencies, international authorities are available for implementation of advance technologies. Similarly, there are positive changes in education, work infrastructure resources, and work meaning. Industry 4.0 increases the demand for employees with competence in mechatronics. Industry 4.0 would create new kinds of jobs. New jobs may shift from the full-time employment model to non-routine jobs (i.e., part-time, temporary, on-call, etc.). A renewed training policy would be demanded and new approaches to training may be visible. Organizations will build their human capital on the basis of new skill sets. Opportunities for more work flexibility and better work-life balance and lowering cultural barriers could arise. In Industry 4.0 humans are more important than ever. Job transformation instead of job losses can be witnessed. New roles will demand more complex tasks. Know-how will be the determining factor at the market. There would be significant decrease in lower skilled jobs.

10.2   Negative Impacts

A high level of digital skills is required for survival in Industry 4.0. Automation will have a negative impact on white-collar jobs in multiple areas from HR to Finance. It would impact health care, education, and law firms. HR departments would be streamlined as most of the white-collar jobs will be replaced by AI. Demand for new skills is an area of potential threat for implementation of Industry 4.0. In addition, training the existing workforce in a specific skill set is also a challenge. On the other hand, motivation, training, retraining skilling, has a cost. Similarly, higher level of employee adaptation to new order skills is important for the success of Industry 4.0. A potential unstable shift in the labor market is another threat. Greater automation will lead to displacement of some of the often low-skilled laborers who perform simple, repetitive tasks. Automation will no longer be restricted to manual, the risky, or dull jobs, but may put at risk many white-collar jobs. Various job positions will become redundant, and some will be newly created. Initially, retraining and qualifications will be a problem factor. In subsequent stages, more educated and skilled workforce in the field of computing, self-learning, and innovation would be in demand. Also, the need for new curricula and disciplines within tertiary education would be required.

11   Conclusion

Thus, a thorough analysis of Industry 4.0 from human resource perspective highlights many important aspects and points out to a fact that technology is a double-edged sword. Its consequences on humans and particularly human work depend upon how we humans embraced it using it as a complementary and sustainable solution rather than perceiving it as a threat to our existence. Use of technology in HR can never be a complete substitute to humans at work. Only need is to further boost our capabilities and cognition to cope up and work hand in hand with this marvellous technology. If still doubtful, ask a HR if she/he would enjoy working with the data on employee's health gathered through wearable sensors or a bunch of papers from hospital.

References

Barman, A., & Das, M. K. *Internet of Things (IoT) as the future smart solution to HRM*.

Cerika, A., & Maksumic, S. (2017). *The effects of new emerging technologies on human resources: Emergence of Industry 4.0, a necessary evil?* Master's thesis, Universitetet i Agder.

Chelliah, J. (2017). Will artificial intelligence usurp White collar jobs? *Human Resource Management International Digest, 25*(3), 1–3.

Christensson, P. (2006). *SMART definition*. Retrieved October 10, 2019, from https://techterms.com.

Chromjakova, F. (2016). Flexible man-man motivation performance management system for Industry 4.0. *International Journal of Management Excelence, 7*(2), 829–840.

Ghobakhloo, M. (2018). The future of manufacturing industry: A strategic roadmap toward Industry 4.0. *Journal of Manufacturing Technology Management*.

Hecklau, F., Galeitzke, M., Flachs, S., & Kohl, H. (2016). Holistic approach for human resource management in Industry 4.0. *Procedia Cirp, 54*, 1–6.

Kagermann, H., Wahlster, W., & Helbig, J. (2013). *Recommendations for implementing the strategic initiative INDUSTRIE*. Acatech–National Academy of Science and Engineering.

Kergroach, S. (2017). Industry 4.0: New challenges and opportunities for the labour market. *Foresight and STI Governance, 11*(4), 6–8. https://doi.org/10.17323/2500-2597.2017.4.6.8.

Li, G., Hou, Y., & Wu, A. (2017). Fourth Industrial Revolution: Technological drivers, impacts and coping methods. *Chinese Geographical Science, 27*(4), 626–637. https://doi.org/10.1007/s11769-017-0890-x.

Lu, Y. (2017). Industry 4.0: A survey on technologies, applications and open research issues. *Journal of Industrial Information Integration, 6*, 1–10.

Muhuri, P. K., Shukla, A. K., & Abraham, A. (2019). Industry 4.0: A bibliometric analysis and detailed overview. *Engineering Applications of Artificial Intelligence, 78*, 218–235.

Müller J. M., Buliga, O., Voigt, K. I. (2018). Fortune favors the prepared: How SMEs approach business model innovations in Industry 4.0. *Technological Forecasting and Social Change*.

Onik, M. M. H., Miraz, M. H., & Kim, C. S. (2018). A recruitment and human resource management technique using Blockchain technology for industry 4.0. In *Proceeding of smart cities symposium* (SCS-2018, pp. 11–16). IET

Piccarozzi, M., Aquilani, B., & Gatti, C. (2018). Industry 4.0 in management studies: A systematic literature review. *Sustainability, 10*(10), 3821.

Qureshi, M. O., & Syed, R. S. (2014). The impact of robotics on employment and motivation of employees in the service sector, with special reference to health care. *Safety and Health at Work, 5*(4), 198–202.

Rao, S. K., & Prasad, R. (2018). Impact of 5G technologies on industry 4.0. *Wireless Personal Communications, 100*(1).

Rosas-Daniel, J. A., Rodríguez-Elias, O. M., Velazquez-Mendoza, M. D. J., & Rose-Gómez, C. E. (2014). A literature review on the use of soft computing in support of human resource management. *Research in Computing Science, 80,* 107–117.

Schumacher, A., Erol, S., & Sihn, W. (2016). A maturity model for assessing Industry 4.0 readiness and maturity of manufacturing enterprises. *Procedia Cirp, 52,* 161–166.

Shivika, T., & Anand, S. (2018). Human resource management: Machine learning perspective. *IJAPRR, V*(III), 23–28.

Shu, I. T. et al. (2018). An overview of Industry 4.0: Definition, components, and government initiatives. *Journal of Advanced Research in Dynamical and Control Systems.*

Sivathanu, B., & Pillai, R. (2018). Smart HR 4.0–how industry 4.0 is disrupting HR. *Human Resource Management International Digest, 26*(4), 7–11.

Trstenjak, M., & Ćosić, P. (2018, January). Challenges of Human Resources Management with implementation of Industry 4.0. In *IoTsm2018.*

Wang, X. L., Wang, L., Bi, Z., Li, Y. Y., & Xu, Y. (2016). Cloud computing in human resource management (HRM) system for small and medium enterprises (SMEs). *The International Journal of Advanced Manufacturing Technology, 84*(1–4), 485–496.

Talent Development Challenges and Opportunities in the 4th Industrial Revolution: A Boundaryless Career Theory Perspectives

Gertrude I. Hewapathirana and Firas Almasri

1 INTRODUCTION

The emerging digital revolution relies heavily on technological advancements, mechanization, and automation using robots, artificial intelligence, and innovative digital technology named the Fourth Industrial Revolution (4IR). With the dawn of the twenty-first century, there are massive technological changes at local and global levels. It is an imminent trend

G. I. Hewapathirana (✉)
College of Business Administration, Gulf University for Science and Technology, Kuwait City, Kuwait
e-mail: Hewapathirana.g@gust.edu.kw

College of Education and Human Development, University of Minnesota, Twin Cities, Minnesota, USA

F. Almasri
College of Arts and Sciences, Gulf University for Science and Technology, Kuwait City, Kuwait
e-mail: Masri.F@gust.edu.kw

Centre for Education Studies, University of Warwick, Coventry, UK

© The Author(s), under exclusive license to Springer Nature Singapore Pte Ltd. 2021
N. R. Al Mawali et al. (eds.), *Fourth Industrial Revolution and Business Dynamics*, https://doi.org/10.1007/978-981-16-3250-1_15

287

that highly developed countries and organizations will rapidly integrate innovations, digitization, robots, and artificial intelligence into almost all aspects of manufacturing and services, replacing some or most human jobs (Fuldauer, 2019). A higher level of digital innovations would increase the gap between the poor and the rich because organizations with enhanced research and development (R&D) take advantage of technological break-throughs requiring different types of boundaryless talents (WEF, 2018). The changing business landscape creates opportunities and challenges that require quick responses and adaptation at individual and organizational levels (WEF, 2018). Regrettably, there is a tendency for most employees who cannot adapt to organizations' emerging talent requirements would lose their jobs. To sustain in the 4IR, organizations need to change their business processes, manufacturing, and work systems. Thus, organizational structures and work boundaries will become precarious and boundaryless. Increasing hyper-competition results in economic shutdowns, unequal development, social, political, and income inequalities widening the gap between the rich and the poor communities (WEF, 2018). Such dramatic changes require a new genre of boundaryless talents to make use of technological advancements in all sizes of businesses (World Trade Report, 2018). In the 4IR environment, individuals' talents become a critical driver for organizations to increase technological innovations to sustain hyper-competition.

The increasing use of artificial intelligence (AI), robots, and digital technology would change the patterns of manufacturing, work systems that affect global communities (Fuldauer, 2019). A higher level of R&D and innovation further increase global competition. To navigate this volatile and highly competitive work environment, employees require developing a higher level of agility and a new set of talents that can take over and manage advanced technological breakthroughs. For organizations, acquiring, developing, and retaining talents become the most critical aspect, and for individuals, swiftly deskilling, reskilling, and acquiring new talents are indispensable. Chambers et al. (1998) found that a talent shortage would hamper the benefit of a new digital revolution while global giants compete to find rare talents. To sustain in the hyper-competitive global business environment, organizations require nurturing, developing, and training high-potential talents; however, retaining extraordinary talents would be the most challenging task (Al-Qeed et al., 2018). The human capital becomes a vital aspect of survival and growth strategies in the 4IR environment. Locating and recruiting high-performing employees, organizations invariably generate a war for talent due to a shortage of best-fit talents (Kravariti & Johnston, 2019; WEF, 2018; Zhang & Bright, 2012). The increasing literature in recent years highlights talent development and management as a high priority area for organizations worldwide to achieve strategic competitiveness (Gallardo-Gallardo et al., 2013; Iles et al., 2010; Kravariti & Johnston, 2019; Owoeye & Muathe, 2018). Since the 4IR impacts on organizations to change their traditional boundaries to accommodate emerging technological advances and innovations, we believe a boundaryless career theory (Arthur & Rousseau, 1996; Briscoe & Hall, 2006, Eby, 2001; Van Buren, 2003) is an appropriate theoretical foundation to build on our analysis of talent development practices, challenges, and opportunities in the 4IR.

This chapter aims to discuss some of the challenges and opportunities for talent development by using the boundaryless career theoretical perspectives. We focus on two overarching research questions: (1). Are we developing the talents we need for a boundaryless career era? (2). How can we overcome the talent development challenges arising from the highly technologized and competitive business world in the 4IR? The subsequent sections include an analysis of related work and concepts, boundaryless career theory context and its applicability to the 4IR, findings and analysis of challenges and opportunities of talents development, and a discussion followed by future research directions and implications for theory and practice. This chapter brings up an important, timely topic relevant to all sizes of organizations and individuals. The propositions would be useful, especially for practitioners and decision-makers, to initiate effective ways of identifying, cultivating, and managing talents to bring competitive edges for organizations.

2   Literature Review and Analysis of Related Work and Concepts

An analysis of relevant literature from various disciplines is used to build the theoretical argument and the analysis. Researchers draw attention to the increasing hyper-competitive environment in the 4IR as the primary driver that forces organizations to change and focus on dynamic strategic directions. As per the World Economic Forum (WEF, 2018), the emerging 4IR geared by rapid technological advancements creates enormous work opportunities that require high-impact multidimensional and agile talents (G.T.C.I, 2019). Inevitably, goods, services, people, information, knowledge, skills, ideas, and innovations move freely across geographic borders (WEF, 2018, v.5). To enhance adaptability, flexibility and capitalize on many benefits arising in the 4IR, organizations are compelled to deviate from traditional business systems by adopting automation, virtual collaborations, flexibility, and so on (Gallardo-Gallardo et al., 2013). To sustain in an unpredictable and unstable global business landscape and align with changing global business context, organizations alter their traditional organizational work processes, patterns, and roles, thus expanding organizational boundaries requiring radical changes in employee development (Liu & Chen, 2013). For example, with Amazon taking the lead to the online superstore to serve a global community and increasing 3D–5D technology, many organizations began focusing on the global digital market space. Traditional organizations are transforming into virtual and mobile workplaces. Organizations are becoming highly flexible with new, smarter structures, and work systems. Employees who got used to traditional work roles and patterns are expected to adapt to changing organizational needs and strategic directions while willingly accepting new work roles. These changes alter traditional work systems, processes, and roles, creating a demand for a new set of high-impact talented employees who can adapt to instant changes utilizing portable and agile talents (Kravariti & Johnston, 2019). However, the problem is a shortage of high-impact talents and a lack of understanding of what specific talents would fill the talent gap in the 4IR business environment (Russell, 2018).

In contrast, increasing social and economic inequality due to a division between high performers and low performers, a lack of understanding of essential talent requirements, and a mismatch of current talents to the technologically advanced business needs result in underutilization of workforce in many countries. While high performers get opportunities to enhance their performance, the low performers may not receive that much attention and talent development opportunities, leading to social inequality (Gallardo-Gallardo et al., 2017). Likewise, there is an ongoing confusion of understanding the meaning and needs of talent. What talent exactly means? Is it possible to set specific limits in the workplace between different types of talent? These questions and confusions are not new, yet we are still struggling to identify and define talents, thus hindering theory development in talent management practices (Gallardo-Gallardo et al., 2017; Tansley, 2011).

2.1 Boundaryless Career Theory and Its Applicability to the 4IR Environment

The concept of a boundaryless career evolved after the 1990s (Van Buren, 2003). A boundaryless career concept depicts both advantages and disadvantages for employers as well as employees. Expanding career boundaries enables organizational flexibility because of the rapidly changing technology; a hyper-competition demands flexibility and adaptability to the changes. Due to changing organizational boundaries, downsizing, and sudden changes to work systems often result in employee layoffs, career changes (Sullivan, 1999), uncertainty, and a lack of job security. Thus, a lifelong career concept is becoming no longer valid. Individuals need to be equipped with multiple talents to adapt to sudden changes to secure jobs. Jobs become a short term, at will or contractual consultancies.

Organizations are increasingly seeking collaboration with all stakeholders, including suppliers, investors, and consumers, and moving forward with high-speed technological changes to meet the instant market demand. This trend invariably requires a new set of talents and mindsets (Russell, 2018). Increasing hyper-competition demands entrepreneurial mindsets of individuals who value flexibility, adapt to change speed, innovate, and integrate creative and innovative work practices. Continually changing work conditions require individuals to adapt to changing work environments swiftly (Toric, 2007). Hence, all stakeholders must understand the changing nature of talents in demand for the 4IR era and provide opportunities to develop new talents at an individual, organizational, national, and global levels. Such changes act as drivers for individuals to master a set of rare, valuable, and high demanding talents to get them hired.

Organizations are searching for leaders and employees with entrepreneurial talents, agile mindsets, and the ability to become lifelong learners. Emerging trends of millennial workers change the traditional way they see the work. World Economic Forum (WEF, 2018) highlights the need for entrepreneurial mindsets as a priority talent for businesses and governments, while individuals need to develop agile mindsets to become lifelong learners. Consumers are becoming more educated as they demand high-quality products and services; continually changing technology brings new challenges at all levels. Individuals are required to become lifelong learners, but they also need to develop new talents that can meet a higher level of efficiencies, productivity, and new ways of thinking to fit into the demands of high-quality job requirements. The future business world would generate enormous boundaryless career opportunities. The 4IR technological innovations open a new range of livelihoods for best-fit talented employees. At the same time, it creates entirely new tasks such as app development to piloting drones to remotely monitoring patient health to certified care workers, driverless automotive, robots and artificially intelligent machines will take over the commanding of workers role (WEF, 2018, v.5).

Traditional business models are being challenged and disrupted (Bider & Jalali, 2011, 2014). Instead of larger, more extensive business processes, smaller business process cycles will be adopted. Brider and Jalali describe examples that agile approach needs to change business modeling as (a) "the phases of process modeling, IT-system design, and manufacturing are merged into one," and (b) "instead of using one big cycle, a series of smaller development cycles will be used" (Bider & Jalali, 2014, p. 1). These changes to business models require knowledge transformation and a new set of boundaryless talents. Technological advancements and automation enable organizations to reduce their workforce while individuals demand shorter workdays to care for families and free time. Increasing demand for new work roles challenges individuals to develop agile mindsets and become agile learners. Working smarter is critical to meet the emerging market demand (Russell, 2018). Customers are becoming much more knowledgeable as they use mobile devices such as smartphones demanding branded quality goods and services. Organizations are challenged to adapt to new technology to retain consumers quickly. However, many organizations are unprepared to compete in today's market for various reasons; inadequate talent supply becomes the apex of such challenges (Ingham, 2006; Zhang & Bright, 2012).

Digitization and the use of sophisticated technology are taking over traditional work systems and organizational structures. Organizations are transforming and becoming more flexible. Organizational structures are shrinking and moving toward downsizing and mergers to make use of resources efficiently. To succeed in the digitalized yet continually transforming organizations and changing nature of work, agile mindsets (Russell, 2018) and agile business processes are needed (Bider & Jalali, 2014). An agile mindset is defined as a set of attitudes and abilities that can support an agile work environment. Agile business processes are adopting automation and innovative technology to improve efficiency (Bider & Jalali, 2014). Getting work done in small self-organized teams, collaborating with others, and adapting instantly to the shifting marketplace are characteristics of organizational changes that require new talents and mindsets. For example, respect, collaboration, willingness to improve, enthusiasm, adding value and a higher level of ability to adapt and add value, openness to change are few examples to name (Denning, 2019).

Boundaryless career theory is also linked to the concept of a boundaryless organization. Because due to the 4IR, organizations invariably alter their traditional manufacturing, work, and product innovation processes, structures, and strategic directions to cope with the changing consumer demands and technological innovations. A few examples are downsizing, mergers and acquisitions, network organizations, and connections across various industries and geographic locations (Dowd & Kaplan, 2005). These changes continuously alter traditional career boundaries and open new needs for reskilling and deskilling (Mirvis & Hall, 1994). With these changes, individuals tend to focus on less stability, high flexibility, and job tenure are no longer a concern in the boundaryless career era (Van Buren, 2003).

Integration of robotic manufacturing systems and artificial intelligence requires new talents to operate and manage organizations. The businesses that can swiftly adapt to the change, use of information smartly and effectively, understand and satisfy rapidly changing customer expectations would have chances to survive (Russell, 2018). The new world of work requires a boundaryless talent to fit into a boundaryless career era. Only a few studies have been attempted to conceptualize talents that fit this changing nature of organizations and technological advances. However, no study explores the challenges of identifying and cultivating talents that can intersect with the concept of a boundaryless career lens.

2.1.1   Evolution of the Concept of Boundaryless Career

Boundaryless careers are defined in many ways. For example, a boundaryless career "unfolds in multiple employment settings, provides opportunities for personal fulfillment and economic prosperity" (Van Buren, 2003, p. 131). The research found that employees who have a passion for expanding their skills and competencies beyond traditional work boundaries prefer temporary work rather than a lifelong commitment to one organization (Arthur & Rousseau, 1996). On the other hand, employers who desire to eliminate high-paying employees demand the remaining employees do extra work beyond traditionally assigned work boundaries (Van Buren, 2003). Changing career boundaries and boundaryless organizations affect both individual and organizational levels. The boundaryless career approach takes a more externally oriented perspective. For individuals, a boundaryless career brings challenges to maintain employability. For organizations, it generates challenges to provide meaningful and exciting work and a wide range of management support to retain talented employees (Van den Brink & Fruytier, 2013).

Broadening career boundaries creates opportunities as well as problems for individuals and organizations. At an individual level, a lack of opportunities to identify, foster, and master talents in demand to enhance employability is a primary concern. Nevertheless, organizations face challenges locating, attracting, and hiring high-impact talents. Creating and providing interesting and meaningful work at an organizational level and learning support to continuously improve individuals' rare talents are necessary to attract critical talents to achieve organizational goals (Van den Brink & Fruytier, 2013).

The traditional way of defining a career shows set boundaries. For example, a career is seen as the unfolding progression of an individual's work experience over time (Arthur et al., 1989), a career consists of one specific job in one or more organizations; an individual can carry the same job in several organizations during his or her lifetime. Usually, in traditional work systems, an individual's skills, knowledge, and experience are confined to one job type. For example, a Human Resource specialist can move to another organization for the same title or a job. On the contrary, a boundaryless career erodes the traditional work boundaries and often results in layoffs and career changes. A person with multiple talents has the potential to shift from one type of job to a vastly different job that requires a separate set of talents. A boundaryless career concept becomes increasingly popular as against traditional lifelong work and individuals binding to one organization.

2.1.2   An Analysis of Conceptualization of a Boundaryless Career in View of the 4IR

Robinson and Miner (1996, p. 78) defined a boundaryless career as a career that unfolds and unconstraint by clear boundaries around job activities. Some others view a boundaryless career crosses traditional organizational boundary (Briscoe & Hall, 2006; Eby, 2001). Sullivan and Arthur (2006) found that psychological and physical mobility and career changes are the prominent characteristics of a boundaryless career. Physical changes in work patterns and arrangements, on the other hand, lead to transform organizations from their traditional work patterns and systems, thus expanding traditional work boundaries. In defining a boundaryless career, Arthur and Rousseau (1996) found six characteristics. They are a career that (a) moves upward and often separates from employers; (b) marketability outside the present employer; (c) uses external networks and information to connect with new opportunities; (d) move away from traditional organizational hierarchies and career advancements; (e) employees more focus on their families and personal matters over sticking to a long-term career in one organization; (f) employees desire a limitless variety of career opportunities in different organizations. In the boundaryless career concept, employees no longer desire to stick to one organization as they seek lucrative new opportunities outside traditional work boundaries. Arthur and Rousseau (1996) conceptualize the boundaryless career based on individuals' physical and psychological mobility. Hall (2004) states that individuals take charge of managing their careers rather than the organization. Hall refers to this as a protein career. The notion of a protein career characterizes as self-directed, flexible, adaptable, and versatile (Greenhaus et al., 2008, p. 278). Individuals take charge of their self-development to fulfill psychological satisfaction. However, these conceptualizations ignore the essential and unique talents of individuals. Further, these conceptualizations are limited to individual occupations and organizational levels as they are not aligning to the external opportunities and global changes taking place in the dynamic market environments (Toric, 2007). Further, organizations are becoming highly flexible and change old business practices to lucrative, high-potential profitable businesses (Greenhaus et al., 2008). Individual talents soon become outdated as a result of such developments. Organizations are reluctant to establish long-term employment relationships and shift hiring rare short-term talents to accommodate instant changes. Traditional job discontinuity, obsolete skill sets are outcomes of such changes. Due to obsolete skill sets and changing market behavior, individuals face job losses (Greenhaus & Parasuraman, 1999). Individuals experience a lack of career opportunities, long-term job security, and employability due to organizational changes (Sullivan, 1999). Such circumstances force individuals to develop more portable skills and essential talents. Continuous self-directed talent development becomes an alternate solution in the boundaryless career concept.

Expanding the previous notion, Sullivan and Arthur (2006) criticize that physical mobility is more comfortable to measure, but psychological movements are not easy to measure. Most studies (Briscoe et al., 2006) limit their focus on physical mobility (movements between organizations, jobs/careers, countries) rather than exploring psychological changes happening within an individual. Further measuring psychological mobility is complicated and blurry from person to person. Based on these findings, Sullivan and Arthur (2006) proposed a two-dimensional quadrant to depict the relationship of psychological and physical mobility as the primary concepts of a boundaryless career. Their argument highlights that an individuals' psychological and physical mobility can be measured based on a degree of mobility of careers and organizational changes. Their research mainly focuses on conceptualizing and operationalizing the relationship between physical and psychological mobility rather than understanding talent development prospects or defining the meaning of talent.

The major characteristics of boundaryless careers are high mobility across geographic boundaries, occupations, and industries. Job security and lifelong work systems are changing (Stone & Deadrick, 2015). In a boundaryless career, individuals define their career based on their occupation and self-identity; thus, organizational identity is no longer critical (Dowd & Kaplan, 2005). Career identity is based on persons' unique talents because individuals take charge of their growth and adapt to changing boundaryless career environments (Mirvis & Hall, 1994).

In the boundaryless career concept, job security, continuum, longer-term benefits, and career growth in one career, and one organization are not applicable as individuals' psychological satisfaction, and expectations become key determinants. For individuals' meaning making out of a job, self, family and career fit, adaptability and mobility of knowledge, skills, and personal aspirations are important determinants in line with the boundaryless career concept. Such changes in organizations and individuals invariably generate many ambiguities to understand what specific talents, when, where, and how to develop best-fit talents that can be highly mobile and adaptable to many organizations or careers. Though Sullivan and Arthur's (2006) conceptualization advances our understanding of major constituents of a boundaryless career, there can be many factors that need to be considered to understand what exact talents would benefit in today's global workplace, especially 4IR.

DeFillippi and Arthur (1996) suggest three ways to learn about career competencies that facilitate a boundaryless career at an individual level. Measuring an individual's motivation and identity will lead to developing specific competencies (Knowing-why) and gathering information. Desired skills and expertise (knowing-how) and assessing their relationship and individual's reputation to understand who needs to be trained (knowing-whom) (see Greenhaus et al., 2008, p. 280) are a few ways to understand the potential of developing specific talents. Gathering information about why, whom, and how questions would enhance our understanding of what that person would bring to the workplace at an individual level. This is limited to an understanding of individuals and it may not address what talents we need to develop. Only those who have a greater level of specific talents and competencies will have greater career mobility in the 4IR business environment, thus leaving the majority unemployed.

Despite many controversies about a boundaryless career conceptualization introduced by many researchers, the following key constructs are still valid in the 4IR environment. Even though employees tend to be flexible, flexible careers require continual updating and mastering new skills and essential talents to remain employable in the 4IR. To sustain the emerging innovative technological advancements, individuals have significant responsibilities. For example, maintain personal networks, upgrade current skills while mastering new skills continually. Arthur and Rousseau (1996) suggest self-enhancements of multiple skills is essential to sustain in the changing global technologized world. Individuals desire their careers to be "personally fulfilling (and perhaps financially lucrative) than single-organizational career" (Van Buren, 2003, p. 133). Based on the above theoretical arguments and conceptual constructs, it is safe to envision some or most of the changing career boundaries would be further widened in today's 4IR flatform. Thus, boundaryless career theory becomes an appropriate podium to explore new talent's requirements in the 4IR global business environment. To establish our argument that a boundaryless career is an imperative foundation in the 4IR, we summarized key characteristics in Table 1.

Especially in the emerging 4IR, individuals need to be flexible and agile, self-directed, self-managed, self-controlled, intrinsically motivated, highly creative, and imaginary thinkers. They look for emotionally rewarding work (Musial, 2015). Individuals seek relationships and networks as sources of locating career opportunities rather than going upward career ladder in one organization. Thus, cultivating networks and

Table 1 Key characteristics of a boundaryless career

Physical mobility	Psychological mobility	Talent development efforts at the individual level
• Frequent career change • Flexibility to move from one career to another • Ability to cross career boundaries • Short term careers • Inter-organizational mobility	• Aim to satisfy self-goals over organizational goals • Relationship and network orientation • Self-identity over organizational identity • The self-control of mastery of talents irrespective of organizational needs • Emotionally rewarding • Self-credibility over organizational credibility	• Self-directed talent development • Self-oriented creativity and innovation • An effort to develop self-credibility, reputation, and self-brand

Source Compiled from Musial (2015), Liu and Chen (2013), Sullivan and Arthur (2006)

building relationships, a higher level of creativity and extrovert thinking, and self-directed talent enhancements are fundamental characteristics. Such individuals seek creative companies with autonomy and freedom to balance family, work, and leisure times. They have the unwritten obligations that make them willingly work extra hours as needed and spend more time socializing outside their work (Musial, 2015). Such employees have their vision to fulfill within the company as well as outside of their organizations. Creative individuals' behaviors invariably match creative organizations; thus, such changes act as drivers for organizations to further expand their work boundaries to attract high-performing employees in the 4IR environment. Especially, psychological, social, and emotional competencies are imperative talents for future career success. Further, boundaryless career theory guides human resource and talent development practitioners to understand some of the portable talents that can help individuals undertake future careers. However, it is still inadequate to understand what portable talents would benefit individuals; hence further studies are essential.

3 RESEARCH METHODOLOGY

To explore the potential changes in the 4IR business world, talent development, and management challenges, we used relevant literature analysis to find answers to the two overarching questions. The databases such as World cat, ProQuest, Google Scholar have been used to find peer-reviewed and related literature. Thus, we used over 100 sources, including peer-reviewed articles, books, magazines, and online research publications, to compile relevant information. The next sections discuss the findings for each question, followed by the discussion.

4 RESULTS AND DISCUSSION

This section discusses the findings to the two overarching research questions and continues the ongoing arguments by many researchers. Q1. Are we developing the talents that are vital for a boundaryless career era? And Q2. How can we overcome the talent development challenges arising from the highly technologized and competitive business world in the 4IR?

4.1 Are We Developing the Talents That Are Vital for a Boundaryless Career Era?

4.1.1 Talents, the Critical Ingredient in the 4IR

Talent is the most vital resource for all organizations to sustain in the 4IR business environment. The 4IR digital revolution induces unprecedented changes in technology and innovations, global economic systems, organizations, and consumer markets. Such changes create a high demand for high-potential talents for organizations. Thus, identifying, developing, and retaining a talented workforce to reap competitive edges becomes the most critical challenge for organizations worldwide (Iles et al., 2010). The increasing intensity of competition, speed of change, emerging smart work systems, virtual marketplaces, innovative products, and services create a necessity to take immediate action to understand specific talent needs in the 4IR changing business landscape. To successfully navigate and sustain in the speed of change and changing work systems, organizations compelled to expand their traditional boundaries and venture into profitable and high demanding products and services. Thus, recruiting high-performing best-fit talents to achieve organizational excellence becomes the primary strategic goal and a grave necessity for organizations to sustain in the 4IR business world (Michaels et al., 2001). However, due to a lack of consensus of various definitions and theoretical clarity, talent management effort is still lag (Ariss et al., 2014; Lewis & Heckman, 2006; Scullion et al., 2010).

4.1.2   What Is the Meaning of Talent?

Increasing discussions on defining and understanding the meaning of talent have gained momentum since the early 1990s. Many researchers proposed diverse definitions; however, the meaning of talent is still ambiguous and unclear (Ariss et al., 2014; Gallardo-Gallardo et al., 2013; Tansley, 2011). Many practitioners and researchers still struggle to answer many questions such as "what is the meaning of talent" and how to identify and develop high demanding talents to fill the increasing shortage of talents worldwide? (Tansley, 2011). Yet, we are still arguing and trying to define the meaning of talent (Gallardo-Gallardo et al., 2013). A clear definition is critical for identifying, nurturing, and disseminating impactful talents to reduce the increasing talent gap, especially in the 4IR (Ariss et al., 2014; Lewis & Heckman, 2006; Scullion et al., 2010). While talent is understood as a critical need, researchers argue that it is difficult to identify, cultivate, and develop talents due to a lack of shared global understanding and confusing definitions of talents (Bider & Jalali, 2014; Nijssen & Paauwe, 2012).

4.1.3   The Dynamic Nature of Future Careers

The 4IR creates a new, more complex, and dynamic global economic environment. Organizations perpetually become global enterprises filled with highly diverse employees due to the mobility of a talented workforce. Skills migration is imperative to capitalize on the emerging opportunities from the 4IR (Schuler et al., 2011). With the rapidly increasing changes to organizational boundaries, careers become nonlinear and boundaryless. The traditional career with linear patterns will change to nonlinear progression, and continuous changes are unavoidable prominent trends (Bergmann & Mendenhall, 2002). Future careers demand multidimensional and more complicated talents and go beyond institutions and traditional occupational boundaries (ibid.). The organization's capability is highly dependent on the capabilities of individuals' talents and their abilities to integrate into the work in their organizations.

4.1.4   Future Demand for Talents

With the transformation of business strategies to benefit from increasingly changing technological advancements in the 4IR, organizations require a new skill set. Hiring managers often find a shortage of best-fit talents. The global talent 2021 report shows that 40 million people in the highly industrialized world are unemployed. In contrast, from the organizational perspective, managers complain about a lack of employees who have the right talents. To meet the increasing demand for talent, a higher level of critical reskilling of the global workforce is essential to fill the talent gap arising from the highly digitized and interconnected world specific to the 4IR era. Technical capabilities and digital business skills have been the most considerable shortage in the Asia–Pacific region due to booming e-commerce and leapfrog strategies (Global Talent Survey, 2021). In Europe, skills to use business software and systems and build internal efficiencies are high priority demands for talents. India has the college-educated largest talent pool that can be a source of talent supply. The global talent report 2021 predicts there is a high-potential talent pool coming from developing countries such as Brazil, China, India, Indonesia, Mexico, and Turkey to the developed world. However, there is a mismatch of demand and supply of talents, mostly due to unawareness of the sources of talents and the type of vital talents.

Organizations require a high-impact talent to take over a new work genre emerging from the 4IR digital revolution. The Global Talent 2021 survey conducted by Oxford Economics identified four critical areas as high demanding skills in the future. They are digital skills, agile thinking skills, interpersonal and communication skills, and global operating skills (Global Talent Survey, 2021).

a. Digital skills category includes digital business skills (50.6%), ability to work virtually (44.9%), understanding of IT software and systems management (40.1%), digital design skills (35.2%), ability to use social media and web technology (29.3%).

b. Agile thinking and abilities to prepare and work under multiple scenarios, innovative thinking, and the ability to deal with complexity and manage paradoxes are critical needs. The agile thinking abilities category includes the ability to envision and prepare for multiple talent scenarios (54.8%), ability to materialize innovation (46.0%), ability to deal with complexity and ambiguity

(42.9%, ability to manage paradoxes and balance opposing views (40.9%) and ability to see the big picture (15.3%).

c. The interpersonal and communication skills category includes co-creativity and barnstorming (48.3%), relationship building with customers (47.4%), teaming both face-to-face and virtual (44.9%), collaboration (30.4%), and oral and written communication (29.0%).

d. With the expanded markets worldwide, a higher level of personnel working in diverse geographic locations is also increasing. Global operating skills include the ability to manage diverse employees, understanding international markets (45.7%), ability to work in multiple overseas locations (37.5%), foreign language skills (36.15), and cultural sensitivity (31.5%) (Global Talent Survey, 2021). The future of jobs reports 2018 conducted by the World Economic Forum shows the changing nature of job-related skills. The talent trends by 2022 are ranked of importance below (WEF, 2018).

Future skills in demand ranking of priority by 2022:

1. Analytical thinking.
2. Innovation thinking.
3. Active learning and learning strategies.
4. Creativity, originality, and initiative.
5. Technology design and programming.
6. Critical thinking and analysis.
7. Complex problem-solving.
8. Leadership and social influence.
9. Emotional intelligence.
10. Reasoning, problem-solving, and ideation.
11. Systems analysis and evaluation.

Source World Economic Forum (2018).

World economic forum predicted shifting of talents over the years. For example, in 2018, it considered complex problem-solving as the second most essential skill, but in 2022 it is predicted to be decreased to 6th place on importance. The second most important skill in 2022 is predicted as active learning and learning strategies. Some of the skills that were considered as important until 2018 but predicted to become obsolete by

2022 are as follows: The following skills have been predicted to become obsolete by 2022 while above mentioned talents are in demand.

- Manuel dexterity, endurance, and precision.
- Memory, verbal, auditory, and special abilities.
- Management of financial, material resources.
- Technology installation and maintenance.
- Reading writing, math, and active listening.
- Management of personnel.
- Quality control and safety awareness.
- Coordination and time management.
- Visual, auditory, and speech abilities.
- Technology use, monitoring, and controlling.

Source World Economic Forum (2018).

Since skills are considered part of a more extensive definition of a talent umbrella, HR practitioners need to identify specific talents in demand to prepare a future workforce. Therefore, the definition of talents become indispensable in identifying and developing a new set of talents that can benefit organizations in the emerging technological advances in the 4IR.

In summary, as per the literature analysis, there is a huge shortage of talents (WEF, 2018). Since the 4IR impact on organizational changes that require a multitude of new skills, the boundaryless career theory can be used to generate a shared understanding of future talent needs by country, industry, or commercial sectors. Further, there should be a venue to publicize emerging talents' requirements so that global organizations can easily locate them while educational institutions can develop new skill development initiatives.

4.2   How Can We Overcome the Talent Development Challenges Arising from the Highly Technologized and Competitive World of Business in the 4IR?

4.2.1   An Overview of Talent Development and Management Practices

The 4IR requires a highly mobile and agile talent to engage in gainful employment in the boundaryless organizations to benefit both parties:

individuals and organizations. Human resourcepractitioners have diffi-culty understanding the meaning of talent and the type of talent needs of various stakeholders (Gallardo-Gallardo et al., 2013). There is an esca-lating debate why cannot define talent to support talent development practitioners to augment the supply of best-fit talents for organizations while individuals can initiate an effort to acquire the most promising talents to fit into dynamic global organizations? (Gallardo-Gallardo et al., 2013; Kravariti & Johnston, 2019; Tansley, 2011). We highlighted above that a lack of specific meaning and a commonly agreeable definition of talent hamper talent development and management efforts (Gallardo-Gallardo et al., 2013; Kravariti & Johnston, 2019; Tansley, 2011). A lack of a specific definition hampers human resource management efforts to identify, cultivate, and manage rare, valuable, inimitable, and high-yielding talents to fill the future organizational needs.

Identifying, cultivating, and managing the supply and demand for talent is generally referred to as the human capital engine. Talent manage-ment (TM) is understood as an ordinary human resource (HR) prac-tice and a succession-planning practice or managing talented employees (Lewis & Heckman, 2006). Many researchers agree that there is very little understanding of TM needs in the future and what specific talents need to be developed (Ariss et al., 2014; Tansley, 2011). Today's skilled migra-tion, expatriation (Al Ariss & Crowley-Henry, 2013), and expanding inherited talents across various generations (Meister & Willyerd, 2010) further constrain a unified definition of talents. Inadequate understanding of future demand for specific talents by nations or organizations is still unresolved research topics that draw many discussions in academia and the business world. Thus, defining talent is a global concern.

The primary concern of ongoing debate is that a lack of consistent talent definition hinders all related talent management efforts. To fill this vacuum, Lewis & Heckman proposed three different ways of inter-preting TM in practice: (1) TM is often used simply as a new term for common HR practices (old wine in new bottles), (2) it can allude to succession-planning practices, or (3) it can refer more generically to the management of talented employees (Lewis & Heckman, 2006, p. 140). Another group of researchers proposed developing a talent pool based on high-performing and low-performing classification (Collings & Mellahi, 2009; Joyce & Slocum, 2012; Tansley, 2011). Different researchers define talent as a natural ability and commitment to work (Gallardo-Gallardo et al., 2013). In this natural ability phenomenon of talent, talent's meaning leads to classifying individuals as high and low performers. High-performers are considered high-impact talent, and low performers are considered unimportant or have low or no impactful talents (Gallardo-Gallardo et al., 2017). However, these classifications of talents lead to further confusion and discrimination of individuals' innate abilities, leading to TM efforts' deprivation and fairness.

4.2.2   Talent Management Issues in the Global Context

Defining talent is fundamental to understand what talents are in demand or shortage. The term "talent" has been discussed since Assyrians, Babylonians, Greek, Romans, and other ancient times and attached various meanings such as weight, capital, natural ability, treasure, mental power or abilities, talent is innate giftedness, and so on (Tansley, 2011). Additionally, various countries use the term "talent" and attached multiple meanings. For example, the European meaning of talent refers to masterful outstanding performance and exceptional talents in a specific field. It is also considered innate giftedness linked to extraordinary performance (Tansley, 2011, p. 268). Talent is understood differently in different contexts. Often there is ambiguity about attributes of talents (Lewis & Heckman, 2006). Evidence shows that researchers failed to develop a precise and explicit definition of talent (Gallardo-Gallardo et al., 2013). Table 2 includes a few examples.

Various researchers defined talent in many ways as depicted in Table 2. Tansley et al. (2007) criticize that though the term talent had been in use for centuries yet, there is no working definition to understand what includes talent and what is talent management. A lack of a precise working definition makes it harder for organizations to develop appropriate policies and talent development practices. Individuals face challenges understanding what specific talent would enhance their employability. Further, it makes it harder for practitioners to identify, design, and plan talent development and training interventions (Tansley, 2011).

Defining talent is the foundation for the identification, development, and management of talents. Different organizations define talent based on specific areas, types, and scope of business and strategies, and so on (Ingham, 2006). An organization may consider some talented employees while another organization may consider the same talent, which is not essential. Some employees who demonstrated specific competencies may consider as talented, but many who never had opportunities to demonstrate their hidden talents and potential may consider not talented. Walker

Table 2   Country specific meanings of talent

Source	Country	Meaning of talent	Focus and meaning
Holden and Tansley (2008)	Japan	Ability and skill or accomplishment (p. 268)	Ability and skills
Holden and Tansley (2008)	France	Divine inspiration, to succeed in something, a particular aptitude	Inner drive, aptitude
Holden and Tansley (2008), Ozhegov (1984)	Russia	Outstanding innate qualities, natural gifts	Inherent giftedness, innate ability
Holden and Tansley (2008), Stanisławski (1994)	Polish	Gift, accomplishment, endowments	Giftedness or inherited qualities
Chinese English Dictionary	China	Trust, shared vision, natural ability, natural quality, innate ability	Natural innate ability

(2002) suggests organizations need to provide opportunities to master and grow people's potential to understand their true talents.

On the other hand, Elegbe (2010) criticizes that talent needs to be understood concerning the individual, business, and geographic context because it is highly situation specific. The context of the enterprise is fundamental to define talent. Talent can be observed by demonstrated behavior and those who did not have the opportunity to display behavior.

Certain grouping of the meaning (Table 3) of talent creates marginalization of employees, and it may negatively affect individual development and job performance. Table 4 shows a multitude of definitions and meanings assigned by various researchers and practitioners.

4.3   Meaning of Talents by Domains

Various groups and organizations define and attach multiple meanings to talent. The following is a collection of the terminology used to depict what is talent. Talent is a dynamic mix of many aspects of an individual's innate psychological and cognitive capabilities, skills, knowledge, aptitude, and behavioral competencies, as summarized below. The findings are categorized into four domains of talents in Table 5.

Table 3 Organizational level definitions of talent—examples

Organization	Definition/characteristics/meaning of "Talent"	Field/nature of work/business/
Gordon Ramsay Holdings	Creative flair	Restaurant chef
Google	Idea person, a Challenger, think outside the box	Virtual workplace
PricewaterhouseCoopers	Energy, applied intelligence, ability to make a distinctive difference, willingness to take the challenge,	Service organization
Europe food processing company	Ability to operate in multiple ways, diverse thinking is the most important talent	Manufacturing and trading company
Chartered Institute of Personnel and Development (2007)	Individuals who make the greatest difference demonstrate high potential to make a difference, high performance	Service organization

Source Tansley (2011; Tansley et al. 2007)

5   PROPOSE TALENT CONCEPTUALIZATION

As per the above analysis, talent is an umbrella concept consisting of enormous unobservable inner attributes of an individual. Thus, it is not limited to knowledge, skills, behaviors, and obvious capabilities and competencies. Most talents include invisible inner cognitive, psychological, and relational thinking that shape their way of seeing and doing the work. It is difficult to measure and observe until a person applied and demonstrated outcomes.

Using the Psychological-Contract theory lens, Höglund (2012) states that talents are "employee perceptions of the extent to which talent qualities are rewarded, and the effect of such perceptions on employee-felt obligations to develop skills" (Höglund, 2012, p. 126). This theory highlights how psychological-contract obligations differ among employees who know they are identified as talent, those who know that they are not identified as talent, and those who do not know whether they are identified as talented. This can relate to the unfairness of defining talents, managers' prejudices, biases in identifying talented people, and creating unfairness at the workplace. Effective decision-making in TM. should be tightly linked to the firm's strategy and corporate culture rather than individualize perceptions.

Table 4 Meanings and definitions of talents by researchers

Source	Definitions	The focus and meaning
Buckingham and Vosburgh (2001, p. 21)	Talent refers to a person's recurring patterns of thought, feeling, or behavior that can be productively applied	Emotional, psychological, and behavioral
Buckingham and Vosburgh (2001), Williams (2000) (see Kravariti & Johnston, 2019, p. 6)	Talent refers to human entities that yield higher outputs than the rest of the workforce in a given business setting	Productivity and outcome
Jericó (2001, p. 428)	The implemented capacity of a committed professional or group of professionals achieves superior results in a particular environment and organization	Superior outcome
Michaels et al. (2001, p. xii), Gallardo-Gallardo et al. (2013)	Sum of a person's abilities—his or her innate gifts, skills, knowledge, experience, intelligence, judgment, attitude, character, and drive. It also includes his or her ability to learn and grow	Innate abilities, attitudes, intelligence, inner drive Skills and knowledge
Lewis and Heckman (2006), Silzer and Dowell (2010) (see Kravariti & Johnston, 2019, p. 6)	Talent refers to specific capabilities certain people develop, intending to produce more benefits for their organization	Comparative capabilities way of thinking mindset
Cheese et al. (2007), Tansley et al. (2007)	Talent is the innate capacity individuals hold and which drives them to stand out	Innate capacity, inner drive to do excellence
Ulrich (2007, p. 3)	Talent equals competence [able to do the job] times commitment [willing to do the job] times contribution [finding Meaning and purpose in their work]	Competence, commitment, willingness to do things

(continued)

Table 4 (continued)

Source	Definitions	The focus and meaning
Cheese et al. (2007, p. 46)	Essentially, talent means the total of all the experience, knowledge, skills, and behaviors that a person has and brings to work	Knowledge, experience, skills, and behaviors, achievement orientation
González-Cruz et al. (2012), Gallardo-Gallardo et al. (2013)	A set of competencies that, being developed and applied, allow the person to perform a certain role skillfully	Competencies to do excellence
Bethke-Langenegger (2011, p. 3), Gallardo-Gallardo et al. (2013)	Talent to be one of those workers who ensure the competitiveness and future of a company (as specialist or leader) through his organizational/job-specific qualification and knowledge, social and methodical competencies, and eager to learn or achievement-oriented characteristics	Qualifications, knowledge, competence, eagerness to develop, achievement orientation
Ulrich and Smallwood (2012)	Talent is a combination of specific employee competencies and an innate drive to accomplish certain working tasks that presuppose those capabilities	Competencies and capabilities
Ulrich and Smallwood (2012, p. 60)	Talent = competence [knowledge, skills, and values required for today's and tomorrows' job, right skills, right place, the right job, right time] × commitment [willing to do the job] × contribution [finding meaning and purpose in their job]	Knowledge, skills, commitment, willingness
Dries (2013), Rana et al. (2013)	Giftedness, strengths, self-motivated person who possess a wide range of competencies, knowledge	Innate giftedness, competence, knowledge, and self-motivation

(continued)

Table 4 (continued)

Source	Definitions	The focus and meaning
Njis et al. (2014, p. 182) (see Kravariti & Johnston, 2019, p. 6)	Systematically developed innate abilities of individuals deployed in activities they like to find important and want to invest energy. It enables individuals to perform excellently in one or more human functioning domains, operationalized as performing better than other individuals of the same age or experience, or as performing consistently at their personal best	Innate abilities, ability to perform excellently
Baker et al. (2019)	Talent in the sports field: innate (i.e., originating in biological elements present at birth), multidimensional (i.e., consisting of capacities from a range of broad cognitive, physical, and psychological categories), emergenic (i.e., involving interactions among factors that combine multiplicatively), dynamic (i.e., evolving across developmental time due to interactions with environments and random gene expression), and symbiotic (i.e., cultural and social factors will determine the ultimate value of an individual's talent)	Innate and biological capacities, cognitive, physical, and psychological strength to perform excellently

Sources E.g., Gallardo-Gallardo et al. (2013), Kravariti and Johnston (2019)

As per the extensive literature analysis, it is commonly agreed that there are many challenges of developing a theory that applies to a global context. Many researchers state that it is indeed a difficult task due to differences in systems, policies, cultures, etc. Having analyzed over 100 articles, the following illustration proposed talent as an umbrella term. Multilayered talents are necessary for individuals to be successful in today's boundaryless career environment. Those layers are broadly

Table 5 The four domains of talent

Psychological flexibility/mobility and natural ability domain	*Mastery of knowledge and cognitive domain of talents*
a. Natural ability, versatile	a. Multiple intelligence
b. Flexibility/adaptability/willingness to change/positive thinking and attitudes	b. Diversity of thoughts
c. Inner drive/innate flexibility/innate capacity/innate abilities	c. Out of the box thinkers
d. Agile thinking/out of the box thinking	d. Exceptional mastery of knowledge and cognitive abilities
e. Giftedness	
f. Feeling/Thoughts/desire to excel	
g. Ability to deal with ambiguity, complexity, see a big picture, take the challenge	
h. Balancing opposed views/visualization	
i. Can-do attitudes of people	
j. Mindfulness to be flexible to take over extra work roles	
k. Passion and motivation	
l. Individual strengths	
m. Positive psychology—feelings, patterns of thoughts	
Commitment/creativity and relational domain	*Exceptional skills/competencies/abilities domain*
a. Commitment to achieving excellence	a. Acquired skills
b. Rational commitment	b. Skills of creative nature of flair to create a new experience or new things
c. Initiative and creativity	c. Competencies to do things differently
d. Social relations/network building	d. Ability and potential to do extra work Ability to do things above and beyond
	e. Cultural sensitivity, respect for diversity
	f. Relationship and network orientation
	g. Social and relational competence

Note Authors finings classified into four domains

identified as a talent tree. A tree consists of multicolored leaves, fruits, stems, roots, and so on. A tree needs nurturing, fertilizing, and a suitable climate to grow. Thus, a metaphor of a tree can be suitable to define talents. A tree can grow as far as it fits the environment and gets essential nutrients to grow. Talent needs to grow to have new branches, fruits, and new leaves. Once leaves become old, they naturally go away. In the same manner, talents become obsolete with global business systems, technology, and the digital revolution in today's 4IR.

Gallardo-Gallardo et al. (2013) conceptualize talent as natural ability, mastery, commitment, and fit. Gagne (2000) conceptualizes talent as exceptional abilities and attitudes that result in an individual's above-average performance. Based on the literature review, the authors organized commonly discussed components of talent into four major domains. They are.

1. Psychological flexibility/mobility and natural ability domain of talent.
2. Mastery of knowledge and cognitive domain of talent.
3. Commitment/creativity and relational domain of talent.
4. Exceptional skills/competencies/abilities and behavioral domain of talent.

Thus, talent can be defined as a sum of the above mentioned four domains of an individual. Our findings of the four domains of talents have some similarities to the conceptual constructs of the boundaryless career theory: physical mobility, psychological flexibility, and mobility at individual levels. Likewise, there is much similarity between the proposed four domains of talents that fit the boundaryless career concept.

6 Conclusion

In conclusion, 4IR generates opportunities and challenges in all aspects of business and human lives. In this environment locating, acquiring, developing, and retaining talent are critical challenges for organizations. Individuals need to take the prompt initiative to identify and develop a new set of boundaryless talents. Boundaryless career theory generates opportunities for researchers to rethink various ways of new conceptualizations and theory expansion to guide organizations and individuals to prepare for the demands for a new type of talent sets. Talent is still a vague and evolving concept; it can have multiple meanings in multiple contexts. Such a diversity of conceptualizations and a lack of universal consensus make it difficult for one country to define and adopt a unified talent management and talent development model. This chapter illuminates the meaning of talent from multiple perspectives and propositions for redefining and reconceptualization of talents. The implications are that the findings would be useful for practitioners and educators to redefine and regenerate knowledge and to understand various talent requirements, innovate various methods and ways of talent identification, acquisition, development, nurturing, and retention. For theoreticians, this study would be an eye-opener, and it proposes to rethink redefining and conceptualizing talents in various contexts. The findings would generate an understanding of self-development for individual learners and acquire multiple talents to better prepare for emerging boundless career requirements in the emerging 4IR.

6.1 Implications

6.1.1 Practical Implications

Having a clear definition and understanding of talent requirements would support many stakeholders in talent development ventures. The implications are that the findings would be useful for practitioners and educators to redefine and regenerate knowledge and to understand a broader meaning of talents, innovate various talent management models, methods, and ways of talent identification, acquisition, development, and retention. It is beneficial for educational and talent development organizations to look for a new set of educational incubators to nurture new talents that can have future career opportunities and higher employability for new graduates.

Social implications are that generating a new set of talents at the national or global level would create efficient talent management programs. To have a better impact, it is essential to have a broader definition and an understanding of specific types of new talents to utilize the nations' talents better. Having developed multidimensional talents generates career opportunities for learners, but it can also have the social benefit of proper utilization of talents to benefit society.

6.1.2 Theoretical Implications

For theoreticians, this study can be an eye-opener to redefine and conceptualize talents in various contexts. In the talent development process, the first step should be to have an agreed definition of talent. Theory building can be initiated first by conceptualizing and developing a clear definition of talents. A boundaryless career theory can be used as a foundational theory for reconceptualization to accommodate external changes and new demand for a new set of talents.

6.1.3   Implications for Individuals and Business Organizations

There is an increasing trend in the layoff of employees. For example, in 2014, Microsoft eliminated 18,000 jobs since it acquired Nokia to realign new jobs and create a new organizational culture. Popular employers such as Hewlett Packard (HP) to cut costs did a similar layoff. Such layoffs cost them at least $1.1–1.6 billion in severance and benefits packages. To fill new jobs, many organizations often report a talent shortage. In 2014, 36% of global employers reported a talent shortage. In 2015 this number increased to 49% in talent shortage (Career Builders.com, 2018). Businesses can disappear overnight, and new businesses mushroom while existing organizations rapidly launch and transform into new ventures in the 4IR. In this transition, talents become obsolete, and a new set of high-performing and high impactful talents will have a high demand. Today's rapidly changing organizational boundaries require a new set of talents. Therefore, having a supporting framework to identify, nurture new talents needs high priority.

On the other hand, individuals prefer more flexible and short-term lucrative careers resulting in higher gains. It is estimated 40% of employees do not want to work for organizations, rather individuals prefer to be their bosses. This transition gained momentum of an emerging new generation of micro-entrepreneurs. Individuals would generate an understanding of self-development and acquire multiple talents to better prepare for the emerging boundaryless career requirements. So that individuals can find alternate jobs despite many layoffs. For organizations, this research is useful in developing innovative ways to locate, identify, nurture, and retain high-performing talents to achieve strategic goals.

6.2   Limitations

This study is compiled using available research literature up to January 2020. However, there can be the most recent research we may not have considered due to the time factor.

6.3   Future Research Directions

Talent development and management fields are still at the infant stage, and there is a high potential for future research to expand the field of study further. The proposed talent tree model needs testing, verification, and validation in different contexts. Therefore, researching various countries around the world would enhance our understanding. Different industries, such as the high-tech industry sector, service sector, manufacturing, and commercial establishments, require different talents. Therefore, sector-wise research to identify specific talent needs would be useful. In addition to explicit technical expertise, knowledge, skills, and other competencies, emic talents that are not visible yet essential to sustain in business are needed research. Exploring the highly vulnerable talents to the external environment and internal organizational factors in different situations and industries is another research area.

Methodological innovations and assessment tools that can be used to assess psychological adaptability, flexibility, commitment, degree of autonomy, viewing situations and resolving problems, inspirations, future expectations, and so on are potential research. Internal mental acumen that supports individuals to be more open and accept various ups and downs on continuing the job, political and mental understanding are a few areas to name that can be potential future research. This research can be carried out in individual, organization, nation, or larger global industry sector or specific socio-cultural contexts.

Theory development research to support talent developers are beneficial. Research on emerging boundaryless organizations, careers, and a new genre of talents is a potential research topic that can expand global organizations' boundaries. At the same time, individuals can quickly locate opportunities to get hired. Matching talents to demand is an important research area. Various talent development efforts and methodological approaches can be tested and verified.

REFERENCES

Al Ariss, A., & Crowley-Henry, M. (2013). Self-initiated expatriation and migration in the management literature present theorizations and future research directions. *Career Development International, 18*(1), 78–96. https://doi.org/10.1108/13620431311305962.

Al-Qeed, M. A., Khaddam, A., Al-Azzam, Z. F., & Atieh, K. A. E. F. (2018). The effect of talent management and emotional intelligence on organizational performance: Applied study on pharmaceutical industry in Jordan. *Journal of Business and Retail Management Research, 13*, 1–14. https://doi.org/10. 24052/JBRMR/V13IS01/ART-1.

Ariss, A. A. L., Cascio, W. F., & Paauwe, J. (2014). Talent management: Current theories and future research directions. *Journal of World Business, 49*, 173–179. https://doi.org/10.1016/j.jwb.2013.11.001.

Arthur, M. B., Hall, D., & Lawrence, B. (1989). Generating new directions in career theory: The case for a transdisciplinary approach. In Michael B. Arthur, Douglas T. Hall, & Barbara S. Lawrence (Eds.), *Handbook of career theory*. Cambridge University Press. https://doi.org/10.1017/CBO978051162545 9.003.

Arthur, M. B., & Rousseau, D. M. (Eds.). (1996). *The boundaryless career: A new employment principle for a new organizational era*. Oxford University Press.

Baker, J., Wattie, N., & Schorer, J. (2019). A proposed conceptualization of talent in sport: The first step in a long and winding road. *Psychology of Sports and Exercise, 43*, 27–33.

Bergmann, B. M., & Mendenhall. (2002). Non-linearity and response—Ability: Emergent order in 21st-century careers. *Human Relations, 55*(1), 5–32. https://doi.org/10.1177/0018726702055001604.

Bethke-Langenegger, P., Mahler, P., & Staffelbach, B. (2011). Effectiveness of talent management strategies. *European Journal of International Management, 5*(5), 524–539. https://doi.org/10.1504/EJIM.2011.042177.

Bider, I., & Jalali, A. (2011). *Agile business process development: Why, how and when applying Chartered Institute of Personnel and Development (2007), Talent management: Strategy, policy, Practice*. Chartered Institute of Personnel and Development, London.

Bider, I., & Jalali, A. (2014). Agile business process development: Why, how and when—Applying Nonaka's theory of knowledge transformation to business process development. *Information Systems and e-Business Management*. https://doi.org/10.1007/s10257-014-0256-1.

Briscoe, J. P., & Hall, D. T. (2006). The interplay of boundaryless and protean careers: Combinations and implications. *Journal of Vocational Behavior, 69*, 4–18. https://doi.org/10.1016/j.jvb.2005.09.002.

Briscoe, J. P., Hall, D. T., & DeMuth, R. L. F. (2006). Protean and boundaryless careers: An empirical exploration. *Journal of Vocational Behavior, 69*, 30–47.

Buckingham, M., & Vosburgh, R. M. (2001). The 21st century human resources function: It's the talent, stupid! *Human Resource Planning, 24*(4), 17–23.

Career Builders.com. (2018). www.careerbuilder.com.

Chambers, E. G., et al. (1998). The war for talent. *McKinsey Quarterly, 3*(3), 44–57.

Cheese, P., Thomas, R. J., & Craig, E. (2007). *The talent powered organization: Strategies for globalization, talent management and high performance.* Kogan Page.

Collings, D. G., & Mellahi, K. (2009). Strategic talent management: A review and research agenda. *Human Resource Management Review, 19*(4), 304.

Denning, S. (2019, August 13). Understanding the agile mindset. *Forbs Magazine.* Retrieved from https://www.forbes.com/sites/stevedenning/2019/08/13/understanding-the-agile-mindset/#158c22205c17.

DeFillippi, R. J., & Arthur, M. B. (1996). Boundaryless contexts and careers: A competency-based perspective. In M. B. Arthur & D. M. Rousseau (Eds.), *The boundaryless career.* Oxford University Press.

Dries, N. (2013). The psychology of talent management: A review and research agenda. *Human Resource Management Review, 23*(4), 272–285. https://doi.org/10.1016/j.hrmr.2013.05.001.

Dowd, K. O., & Kaplan, D. M. (2005). The career life of academics: Boundaried or boundaryless? *Human Relations, 58,* 699–721.

Eby, L. T. (2001). The boundaryless career experiences of mobile spouses in dual-earner marriages. *Group and Organization Management, 26,* 343–368.

Elegbe, J. A. (2010). *Talent management in the developing world adopting a global perspective.* Gower.

Fuldauer, E. (2019). *Robots, AI and automation: The 4th industrial revolution is here.* Retrieved from https://www.smartcitylab.com/blog/digital-transformation/robots-ai-and-automation-the-4th-industrial-revolution-is-here/.

Gagne. F. (2000). Understanding the complete choreography of talent development through DMGT-based analysis. In K. A. Heller, F. J. Monks, R. F. Subotnik, & R. J. Sternberg (Eds.), *International handbook of giftedness and talent.* Elsevier Science.

Gallardo-Gallardo, E., Dries, N., & González-Cruz, T. F. (2013). What is the meaning of 'talent' in the world of work? *Human Resource Management Review, 23*(4), 290–300. https://doi.org/10.1016/j.hrmr.2013.05.002.

Gallardo-Gallardo, E., Thunnissen, M., & Scullion, H. (2017). A contextualized approach to talent management: Advancing the field. *The International Journal of Human Resource Management.* Advance online publication. https://doi.org/10.1080/09585192.2016.1275292.

Global Talent Competitiveness Index (G.T.C.I.). (2019). Retrieved from http://www.adeccogroup.com.

Global Talent Survey. (2021). How the new geography of talent will transform human resource strategies. *Oxford Economic.* Retrieved from https://www.oxfordeconomics.com/Media/Default/Thought%20Leadership/global-talent-2021.pdf.

González-Cruz, R. D., Fonseca, V. C., & Darling, E. M. (2012). Cellular mechanical properties reflect the differentiation potential of adipose-derived mesenchymal stem cells. *Proceedings of the National Academy of Sciences, 109*(24), E1523–E1529.

Greenhaus, J. H., Callanan, G. A., & DiRenzo, M. (2008). A boundaryless perspective on careers. In *The Sage handbook of organizational behavior* (Vol. 1). https://doi.org/10.4135/9781849200448.n16.

Greenhaus, J. H., & Parasuraman, S. (1999). Research on work, family, and gender: Current status and future direction. In G. N. Powel (Ed.), *Handbook of gender and work* (pp. 391–412). Sage.

Hall, D. T. (2004). The protean career: A quarter-century journey. *Journal of Vocational Behavior, 65*, 1–13.

Höglund, M. (2012). Quid pro quo? Examining talent management through the lens of psychological contracts. *Personnel Review, 41*(2), 126–142. https://doi.org/10.1108/00483481211199991.

Holden, N. J., & Tansley, C. (2008, July). *Talent' in European languages: A philological analysis reveals semantic confusions in management discourse.* Paper presented at the Critical Management Studies Conference, Manchester Business School, Manchester.

Iles, P., Preece, D., & Chuai, X. (2010). Talent management as a management fashion in towards a research agenda. *Human Resource Development International, 13*(2), 125–145.

Ingham, J. (2006). Closing the talent management gap: Harnessing your employee's talent to deliver optimum business performance. *Strategic H.R. Review, 5*(3), 20–23.

Jericó, P. (2001). La gestión del talento: Enfoque conceptual y empírico. *Boletín de Estudios Económicos, LVI*(174), 423–441.

Joyce, W. F., & Slocum, J. W. (2012). Top management talent, strategic capabilities, and firm performance. *Organizational Dynamics, 41*(3), 183–193. https://doi.org/10.1016/j.orgdyn.2012.03.001.

Kravariti, F., & Johnston, K. (2019). Talent management: A critical literature review and research agenda for public sector human resource management. *Public Management Review.* https://doi.org/10.1080/14719037.2019.1638439.

Lewis, R. E., & Heckman, R. J. (2006). Talent management: A critical review. *Human Resource Management Review, 16*(2), 139–154. https://doi.org/10.1016/j.hrmr.2006.03.001.

Liu, E., & Chen, P. (2013). The effect of game-based learning on students' learning performance in science learning—A case of "Conveyance Go". *Procedia—Social and Behavioral Sciences, 103*, 1044–1051. https://doi.org/10.1016/j.sbspro.2013.10.430.

Meister, J. C., & Willyerd, K. (2010). Mentoring millennials. *Harvard Business Review, 88*(5), 68–72.

Michaels, E., Handfield-Jones, H., & Axelrod, B. (2001). *The war for talent.* Harvard Business School Press.

Mirvis, P., & Hall, D. (1994). Psychological success and the boundaryless career. *Journal of Organizational Behavior, 15,* 365–380. https://doi.org/10.1002/job.4030150406.

Musial, M. (2015). A conceptual framework for boundaryless careers and their management in creative industries: The creative freedom/control paradox. *Journal of Innovation Economics and Management, 3*(18), 99–118.

Nijs, S., Gallardo-Gallardo, E., Dries, N., & Sels, L. (2014). A multidisciplinary review into the definition, operationalization, and measurement of talent. *Journal of World Business, 49*(2), 180–191.

Nijssen, M., & Paauwe, J. (2012). H.R.M. in turbulent times: How to achieve organizational agility? *The International Journal of Human Resource Management, 23*(16), 3315–3335. https://doi.org/10.1080/09585192.2012.689160.

Owoeye, I., & Muathe, S. M. (2018). Competence-enhancing interventions and organizational performance: A theoretical review. *Journal of Human Resource Management, 6*(2), 67–77. https://doi.org/10.11648/j.jhrm.20180602.14.

Ozhegov, S. I. (Ed.). (1984). *Slovar' russkogo yazyka* [Dictionary of the Russian language]. Russkii Yazyk.

Rana, G., Goel, A., & Rastogi, R. (2013). Talent management: A paradigm shift in Indian public sector. *Strategic H.R. Review, 12*(4), 197–202. https://doi.org/10.1108/SHR-02-2013-0012.

Robinson, D., & Miner (1996). A careers change as organizations learn. In M. Arthur & D. Rousseau (Eds.), *The boundaryless career.* Oxford University Press.

Russell, M. (2018). *A digital-first future requires an agile mindset.* Retrieved from https://www.just-style.com/analysis/a-digital-first-future-req uires-an-agile-mindset_id133537.aspx.

Schuler, R., Jackson, S., & Tarique, I. (2011). Global talent management and global talent challenges: Strategic opportunities for IHRM. *Journal of World Business, 46,* 506–516. https://doi.org/10.1016/j.jwb.2010.10.011.

Scullion, H., Collings, D., & Caligiuri, P. (2010). Global talent management (global H.R.M.). *Journal of World Business, 45*(2), 105–108. https://doi.org/10.1016/j.jwb.2009.09.011.

Silzer, R. F., & Dowell, B. E. (2010). *Strategy-driven talent management: A leadership imperative.* Jossey-Bass.

Stanisławski, J. (Ed.). (1994). *The great polish-english dictionary.* Wiedza Powszechna.

Stone, D. L., & Deadrick, D. L. (2015). Challenges and opportunities affecting the future of human resource management. *Human Resource Management Review, 25*(2), 139–145.

Sullivan, S. E. (1999). The changing nature of careers: A review and research agenda. *Journal of Management, 25*, 457–484.

Sullivan, S., & Arthur, M. (2006). The evolution of the boundaryless career concept: Examining physical and psychological mobility. *Journal of Vocational Behaviour, 69*, 19–29. https://doi.org/10.1016/j.jvb.2005.09.001.

Tansley, C. (2011). What do we mean by the term "talent" in talent management? *Industrial and Commercial Training, 43*(5), 266–274. https://doi.org/10.1108/00197851111145853.

Tansley, C., Turner, P. A., Foster, C., Harris, L. M., Stewart, J., Sempik, A., & Williams, H. (2007). *Talent: Strategy, management and measurement, research into practice.* CIPD.

Thorne, K., & Pellant, A. (2006). *The essential guide to managing talent: How top companies recruit, train and retain the best employees.* Kogan Page.

Toric, B. (2007). Boundaryless career—Implications for individual and organizational learning. *GRIN Verlag.* https://www.grin.com/document/112909.

Ulrich, D. (2007). The talent trifecta. *Workforce Management, 86*(15), 32–33.

Ulrich, D., & Smallwood, N. (2012). What is talent? *Leader to Leader, 63*, 55–61. https://doi.org/10.1002/ltl.20011.

Van Buren, H. J. (2003). Boundaryless careers and employability obligations. *Business Ethics Quarterly, 13*(2), 131–149.

Van den Brink, M., & Fruytier, B. (2013). Talent management in academia: Performance systems and HRM practices. *Human Resource Management Journal, 23*(2), 180–195. https://doi.org/10.1111/j.1748-8583.2012.00196.x.

Walker, J. W. (2002). Perspectives: Talent pools: The best and the rest. *Human Resource Planning, 25*(3), 12–14.

WEF. (2018). *World Economic Forum.* Retrieved from http://www3.weforum.org/docs/WEF_AM19_Meeting_Overview.pdf.

Williams, M. (2000). *The war for talent: Getting the best from the best.* Chartered Institute of Personnel and Development (CIPD).

World Economic Forum. (2018). *The future of jobs report.* Center for the new economy and Society. Retrieved from http://www3.weforum.org/docs/WEF_Future_of_Jobs_2018.pdf.

WTO. (2018). *The future of world trade: How digital technologies are transforming global commerce.* Geneva. Retrieved from https://www.wto.org/english/res_e/publications_e/world_trade_report18_e.pdf.

Zhang, S., & Bright, D. (2012). Talent definition and talent management recognition in Chinse private-owned enterprises. *Journal of Chinese Entrepreneurship, 4*(2), 143–163. https://doi.org/10.1108/17561391211242753.

INDEX